Routledge Revivals

Environmental Quality Manag

Environmental Quality Management provides a quantitative analysis of regional residuals environmental quality management in the Lower Delaware Valley. Originally published in 1976, this study takes a management outlook to discuss new systems such as a non-linear aquatic eco-system model and reaches conclusions which have influenced research and management decisions about REQM across the world. This title will be of interest to students of Environmental Studies.

Environmental Quality Management

An Application to the Lower Delaware Valley

Walter O. Spofford, Jr., Clifford S. Russell and
Robert A. Kelly

RFF PRESS
RESOURCES FOR THE FUTURE

First published in 1976
by Resources for the Future Inc.

This edition first published in 2016 by Routledge
2 Park Square, Milton Park, Abingdon, Oxon, OX14 4RN
and by Routledge
711 Third Avenue, New York, NY 10017

Routledge is an imprint of the Taylor & Francis Group, an informa business

Publisher's Note
The publisher has gone to great lengths to ensure the quality of this reprint but points out that some imperfections in the original copies may be apparent.

Disclaimer
The publisher has made every effort to trace copyright holders and welcomes correspondence from those they have been unable to contact.

A Library of Congress record exists under LC control number: 76047138

ISBN 13: 978-1-138-95567-7 (hbk)
ISBN 13: 978-1-315-66615-0 (ebk)
ISBN 13: 978-1-138-95569-1 (pbk)

Environmental quality management / *an application to the Lower Delaware Valley*

WALTER O. SPOFFORD, JR., CLIFFORD S. RUSSELL,
and ROBERT A. KELLY

RESEARCH PAPER R-1

RESOURCES FOR THE FUTURE / WASHINGTON, D.C.

Resources for the Future is a nonprofit organization for research and education in the development, conservation, and use of natural resources and the improvement of the quality of the environment. It was established in 1952 with the cooperation of the Ford Foundation. Part of the work of Resources for the Future is carried out by its resident staff; part is supported by grants to universities and other nonprofit organizations. Unless otherwise stated, interpretations and conclusions in RFF publications are those of the authors; the organization takes responsibility for the selection of significant subjects for study, the competence of the researchers, and their freedom of inquiry. This book is a product of RFF's Quality of the Environment Division, directed by Walter O. Spofford, Jr.

Library of Congress Catalog Card Number 76-47138
ISBN 0-8018-1937-7

Manufactured in the United States of America

Published September 1976. $5.75.

Table of Contents

List of Tables

List of Figures

Foreword

The research reported in this paper represents a pioneering effort in quantitative analysis of, and a major output from research at RFF on, regional residuals-environmental quality management. Research in this area at RFF evolved naturally from analyses of a single environmental medium, such as water quality management and air quality management. The analytical framework and methodology evolved over time, stimulated by efforts to develop residuals-environmental quality management (REQM) strategies in various regions, both in the United States and abroad, in some of which efforts RFF staff was directly involved. In the United States, the Clean Air Act Amendments of 1970, mandating the preparation of state implementation and air quality control region plans for air quality management, and the Federal Water Pollution Control Act Amendments of 1972, mandating the preparation of areawide waste treatment and river basin water quality management (208) plans, added further impetus to the development of analytical methodology for REQM. In addition, many state and local jurisdictions in the United States have been, and are, involved in preparing single-medium or, in a few cases, multimedia REQM plans.

There are at least four important points about this research effort which should be stressed. First, this study represents one of the very few quantitative analyses of integrated residuals management with a management orientation. That the effort was accomplished at all is a major tribute to the authors. Second, the study represents a pioneering effort to explore the problems of, and benefits from, incorporating a nonlinear aquatic ecosystem model explicitly within the framework of an optimization analysis for management decisions. Third, this research, in conjunction with previous research on residuals management at RFF, has already had substantial influence on both research and management decisions in various parts of the world. Examples include regional or national analyses or both in Australia, Czechoslovakia, France, Federal Republic of Germany, Mexico, Poland, Switzerland, and Yugoslavia. Similarly in the United States, studies at various scales and in various places have been influenced by the research on the Lower Delaware Valley. Fourth, this effort represents the importance and usefulness of, and payoffs from, interactions between researchers and those involved in actually trying to develop management plans for particular regions. Almost from the beginning of this effort, consulting assignments undertaken by various members of the group involved in the research enabled making inputs into specific management contexts and developing insights into the kinds of problems which those on the "firing line" face in all societies. Such experience in our view has had a very salutory effect on the research itself and, in addition, comprises an important mode of "transferring" research results.

As discussions of incorporating air quality management and solid residuals management within the scope of areawide waste treatment management (208) plans proceed in the United States, the utility of the Lower Delaware Valley study should become even more apparent. Although many questions have been tackled and some answered, there still remains fruitful ground to be plowed, both by the authors of this report and by others.

March 1976 Blair T. Bower and Allen V. Kneese

Preface

Concerns with air and water quality, and the problems of solid wastes disposal, are not unique to our modern, affluent society. Some of these concerns and problems go far back in history. But the bulk of the modern-day environmental quality problems really began with the industrial revolution during the latter part of the eighteenth century. Sanitary conditions in the urban areas grew steadily worse beginning with this period, as did air and water quality. In most cities, air quality was substantially worse than it is today due to the extensive burning of "soft" coal. Sewage and solid wastes posed difficult handling and disposal problems for the urban dwellers, and poor water quality resulted in serious public health problems for the people of those times.

The first quarter of the twentieth century witnessed major improvements in the ambient environmental quality of the urban areas in this country. Air and water quality and wastewater and solid wastes disposal problems were beginning to be attacked both scientifically and extensively, primarily under the impetus of the major health-related impacts of the degraded environment. State and local public health departments played major roles in these efforts. Major technical advances occurred in the provision and treatment of water for potable and related uses and in the treatment of municipal wastewaters. Approaches to waste disposal problems were systematic, effective, and certainly advanced for their times. Nevertheless, air quality, water quality, and solid wastes disposal were all dealt with separately, with very little consideration of interactions among the three sectors. Separate consideration, and management, of the three environmental media prevailed in most areas of the country and at all levels of government until the last decade or so.

About ten years ago, explicit recognition of, and concern with, the linkages and interrelationships among forms of residuals (wastes) and among the environmental media--air, water, and land--began to grow. It had become obvious that measures taken to improve air quality sometimes exacerbated water quality problems, and vice versa, and that measures taken to improve air or water quality usually resulted in the generation of increased quantities of solid residuals that had to be handled and disposed of. In some situations, these quantities were becoming large enough, and the disposal problems difficult enough, so that increased attention was directed toward reuse and recycling. This attention was not motivated by the scarcity of raw materials, as was the case during World War II, but by the desire to reduce the steadily increasing quantities of residuals that had to be handled and disposed of in some manner by our high consumption society.

Among the first social scientists to become concerned with these intermedia linkages were Allen Kneese and Blair Bower, director and associate director, respectively, of the Quality of the Environment Program at Resources for the Future. Both had made explicit attempts to incorporate these linkages in their analyses. Allen Kneese modified and extended general equilibrium economic models by introducing a materials balance approach, and Blair Bower organized and directed an empirical study of the New York City region that featured the importance of these linkages in residuals management. Based on the results of these two studies, as well as on earlier industry studies conducted at RFF, Allen Kneese and Blair Bower brought two of the authors of this report, Walter Spofford and Clifford Russell, to RFF in 1968 to develop a set of quantitative techniques and methodologies for analyzing alternative residuals management strategies that would reflect the linkages and tradeoffs among the three environmental media. It was intended from the start that the approach would be quantitative and comprehensive.

In the beginning of the effort there were no bounds placed on the methodologies to be used or on the spatial and time dimensions of the analysis. However, it soon became evident during the methodological development phase that a quantitative analysis of the intermedia linkages would require us to focus on a single region. To incorporate information on the environmental impacts of various management strategies, we decided that the regional model should be designed to allow the use of both linear and nonlinear air and water quality models, including aquatic ecosystem models. Finally, we decided that the regional model should be structured as an optimization management model as opposed to a recursive simulation management model because of the ability of the former to provide directly the least cost solution for any given set of conditions. Thus, in the beginning, our two major research objectives were: (1) an investigation of the importance of including the linkages among forms of residuals, and among the three environmental media, in the analysis of alternative residuals management strategies for a given region; and (2) an investigation of the potential benefits and the problems of including nonlinear ecosystem models within an optimizing type management model.

The first phase of the research involved the development of a methodology for analyzing the impacts on aggregate (regional) costs, and on regional ambient environmental quality, of alternative residuals management strategies. To illustrate the newly developed methodology, which is an extension of, but based on, the earlier water quality management models, it was applied for selected residuals to a hypothetical region with a relatively small number of residuals sources. About the time we had finished the first version of the regional model, and had applied it to a hypothetical region, we began working with Edwin Haefele, a political theorist and colleague of ours at RFF, who was pioneering the development of legislative bargaining and vote-trading models. Haefele was concerned with the ways in which decisions on the levels of use of public goods were made in the United States. In particular, he was concerned with the processes available to society for selecting desired levels of environmental quality, the distributions of ambient environmental quality, and the distributions of the costs of achieving and maintaining the different levels of ambient environmental quality. In earlier analyses of alternative regional environmental quality management strategies, a major emphasis had been placed on regional efficiency, and very little thought given to the distribution of these costs. Thus, one of the important objectives of Haefele's work was to approach, simultaneously, the two issues of desired levels of environmental quality and of who pays for achieving and maintaining these levels. Another objective was to determine how different representation formulae--representation based on population; one vote per county; one vote per state; and so on--in an environmental quality "shed" (the area within which the management decisions are to be made) would affect decisions concerning both levels of ambient environmental quality and the distribution of costs. The result of Haefele's interest was a major restructuring of the earlier model to provide information on the distribution of costs by political jurisdictions for each distribution of ambient environmental quality. And once again, to illustrate this extension of the regional model, the new version was applied to a hypothetical region. Thus, the third major research objective was to investigate ways of providing, and constraining, cost distributional information in the analysis of alternative regional residuals management strategies.

All this didactic model development was fine, but the usefulness of the approach for policy decision had not yet been demonstrated. An application in the real world was necessary, and the eleven-county Lower Delaware Valley region of Pennsylvania, New Jersey, and Delaware was selected for the illustrative case study. Soon thereafter, in the spring of 1971, our colleague and coauthor, Robert Kelly, joined us at RFF to develop an aquatic ecosystem model of the Delaware Estuary. The regional application, which involved numerous people at RFF, got under way in the spring of 1972.

Previous to the empirical application phase, a number of spinoffs of the regional modeling research effort had emerged. Lack of information on important industrial dischargers in the region, such as petroleum refineries and steel mills, had proved to be a major obstacle. Therefore, in conjunction with an ongoing program of studies of water use and residuals management in industry, a major effort was undertaken at RFF to study the alternative production processes and residuals generation and discharge characteristics of these two industries. Analyses of thermal power generation had already been completed at RFF, and a study of the pulp and paper industry was well under way. The interaction between the industry studies and the regional studies at RFF stimulated increased attention to the economics of reuse and recycling, leading to four subsequent RFF studies: two dealing with the more conceptual aspects of reuse and recycling of materials, and two empirical studies--one on steel scrap and the other on used newspapers and used corrugated containers.

The Lower Delaware Valley regional modeling application took about three years, from 1972 to 1975. A major effort in the first year was placed on the development and calibration of an ecosystem model of the Delaware Estuary and on the continuation of the development of a nonlinear programming algorithm and computer "software" package for application to large-scale regional residuals management problems. In addition, a substantial amount of time was spent during the first year in locating data sources and usable data. The second year was spent primarily on the construction of production and consumption activity models. During the final year, most of the time was spent on modifications and additions to the regional model, on making production runs of the model, and on the analysis of the huge quantity of output available from the regional model. The regional modeling team of Kelly, Russell, and Spofford remained together until the spring of 1975 when Robert Kelly left RFF to assume a position with the Fisheries and Wildlife Department of the Ministry for Conservation in Melbourne, Australia. Clifford Russell and Walter Spofford are still at RFF.

This is a report on the results of the first set of production runs using the regional model with the RFF nonlinear aquatic ecosystem model of the Delaware Estuary. Because of the computational time required due to the nonlinear aspects of the regional model, we have thus far been unable to make detailed analyses of the distributional implications of various strategies or to combine our regional model with Edwin Haefele's political vote-trading model of this region. We are currently in the process of linearizing the regional model by removing the ecosystem model of the estuary and replacing it with a linear dissolved oxygen model. This modification of the regional model will enable us to operate the model in conjunction with the vote-trading model and, thus, to finish the scheduled analyses of alternative residuals management strategies for the region.

This has been a long, detailed, and complex research project, and its success involved a number of people other than the authors of this report. First and foremost, we gratefully acknowledge the continuous support of, and help from, our colleagues Blair Bower and Allen Kneese throughout the duration of this project. The project was originally conceived by them, and without their support and patience it would not have been possible.

We are also appreciative to William Vaughan and Elizabeth Reid for their efforts in assembling data from the Delaware River Basin Commission; to William Vaughan for his development of the large steel production model upon which the five smaller steel mill models in the Lower Delaware Valley model are based; to Marilyn McMillan for her help in developing the module containing the large dischargers of gaseous residuals, which is based on a modification of EPA's Implementation Planning Program; to James Sawyer for his help in constructing the paper plant models, based on analyses of the pulp and paper industry by Blair Bower and consultants George Löf and Mont Hearon; and to Abraham Michaels for his inputs to the development of the municipal solid waste management-paper recycling module.

We are also grateful to Louanne Sawyer and Pathana Thananart for their continuous and essential assistance in the computer programming aspects of the regional application. Louanne Sawyer's help with the linear programming code, the software associated with the nonlinear programming algorithm, and her continued efforts to improve the computational efficiency of the regional model are especially appreciated. Pathana Thananart was of particular help to us in making operational EPA's air dispersion model, in modifying the gaseous emissions inventory data to enable an analysis of various stack aggregation techniques, and in developing the air dispersion matrices that were incorporated in the regional model.

A number of people outside of RFF were also very helpful to us: Ralph Porges, Richard Tortoriello, and Peter K. MacEwen of the Delaware River Basin Commission in Trenton, New Jersey; Robert E. Neligan, Herschel H. Slater, Jerome B. Mersch, Joseph A. Tikvart, and William Cox, all of the U.S. Environmental Protection Agency's Air Pollution Control Office, at the time located in Durham, North Carolina; and Roger A. Smith of the Delaware Valley Regional Planning Commission in Philadelphia.

Finally, we are grateful to Vera Ullrich for her efforts in typing and proofreading the various drafts of this report, as well as the final version, and for her efforts in following the report through to completion and publication.

If we have overlooked anyone, it has certainly been unintentional. This research effort involved many individuals, and it would have been impossible without their help and cooperation. It is an example of the difficulties of, and the positive payoffs from, a truly interdisciplinary team effort. We hope that the quality of the results vindicates their support.

Walter O. Spofford, Jr.

March 1976

Clifford S. Russell

I. Introduction

This paper describes the essential elements of a regional integrated residuals management model developed at Resources for the Future during the last three years by an interdisciplinary team representing the fields of political science, economics, ecology, and engineering.[1] This illustrative application to the 11-county Lower Delaware Valley region represents the final phase--an application in the real world--of a research effort at RFF which has concentrated on the development of regional residuals management models to aid government in establishing public policy on regional environmental quality--air, water, and land. A number of publications describing in detail the various stages in the development of the regional residuals management model are available [1-11]. Published information is also available on some of the component studies used in the development of the Lower Delaware Valley model [12-19].[2]

---

[1] The modeling team consists of the authors of this paper and Edwin T. Haefele, with the assistance of Louanne Sawyer, Pathana Thananart, Blair T. Bower, and James W. Sawyer, Jr. Edwin Haefele developed a legislative bargaining and vote-trading model which is not reported in this paper. However, the management model reported here is designed to operate in conjunction with Haefele's political model.

[2] Unfortunately, many of the more important details of the modeling effort are available only in unpublished RFF reports at this writing. Some of these reports are referred to here. We are, however, currently in the process of writing a final report on the Lower Delaware Valley application which is substantially more inclusive than this paper and which incorporates most of the relevant details.

This summary of the Lower Delaware Valley residuals management study and regional model places emphasis both on a description of the region and the application and on the presentation and analysis of an initial set of information obtained as output of the regional model. This summary does not contain the details of the individual submodels that the regional model comprises. The material presented in this paper is organized as follows: the objectives of the study; a description of the region and its environmental quality problems; the regional residuals management model; management strategies being analyzed with the use of the model; the results achieved to date; special features of the results; policy implications; and the difficulties encountered, research needs, and future research plans.

II. The Objectives of the Study

The illustrative application to the Lower Delaware Valley of the regional residuals management framework developed previously at RFF [1,2] had several objectives involving both the development of methodology and the provision of more and better information upon which to base public policy on matters of regional environmental quality:

1. To investigate the importance of including in a single model the linkages among gaseous, liquid, and solid residuals and among the various environmental media.

2. To explore the computational problems inherent in scaling up from small didactic models to large-scale regional applications (e.g., how best to deal with large amounts of data; how to aggregate without losing too much information; how to decompose management models so that

they become computationally tractable; and how to present the results so that they are useful for informing policy decisions).

3. To explore the feasibility of incorporating within an optimization framework, rather than within the more popular recursive simulation model, complex nonlinear models of the natural world and a large number of residuals management options inherent in any real world regional application so that least-cost ways of achieving various levels of ambient environmental quality could be investigated.

4. To explore ways of designing regional residuals management models to provide information on the distribution of costs and environmental quality that would be useful in a legislative, as well as an executive agency, setting. For this purpose, distributional constraints on environmental quality (air, water, and land) and on consumer costs over geographic units (electricity bills, heating fuels, sewage disposal, solid waste disposal, and regional in-stream aeration) were provided.

5. To present the policy implications of alternative residuals management strategies within a regional context as illustrated by the output of the Lower Delaware Valley residuals management model.

III. Description of the Region and Its Environmental Quality Problems

The region to which we are applying the residuals management modeling framework is an 11-county area of Pennsylvania, New Jersey, and Delaware which is referred to as the Lower Delaware Valley (see Figure 1). This region consists of Bucks, Montgomery, Chester, Delaware, and Philadelphia Counties in Pennsylvania; Mercer, Burlington, Camden, Gloucester,

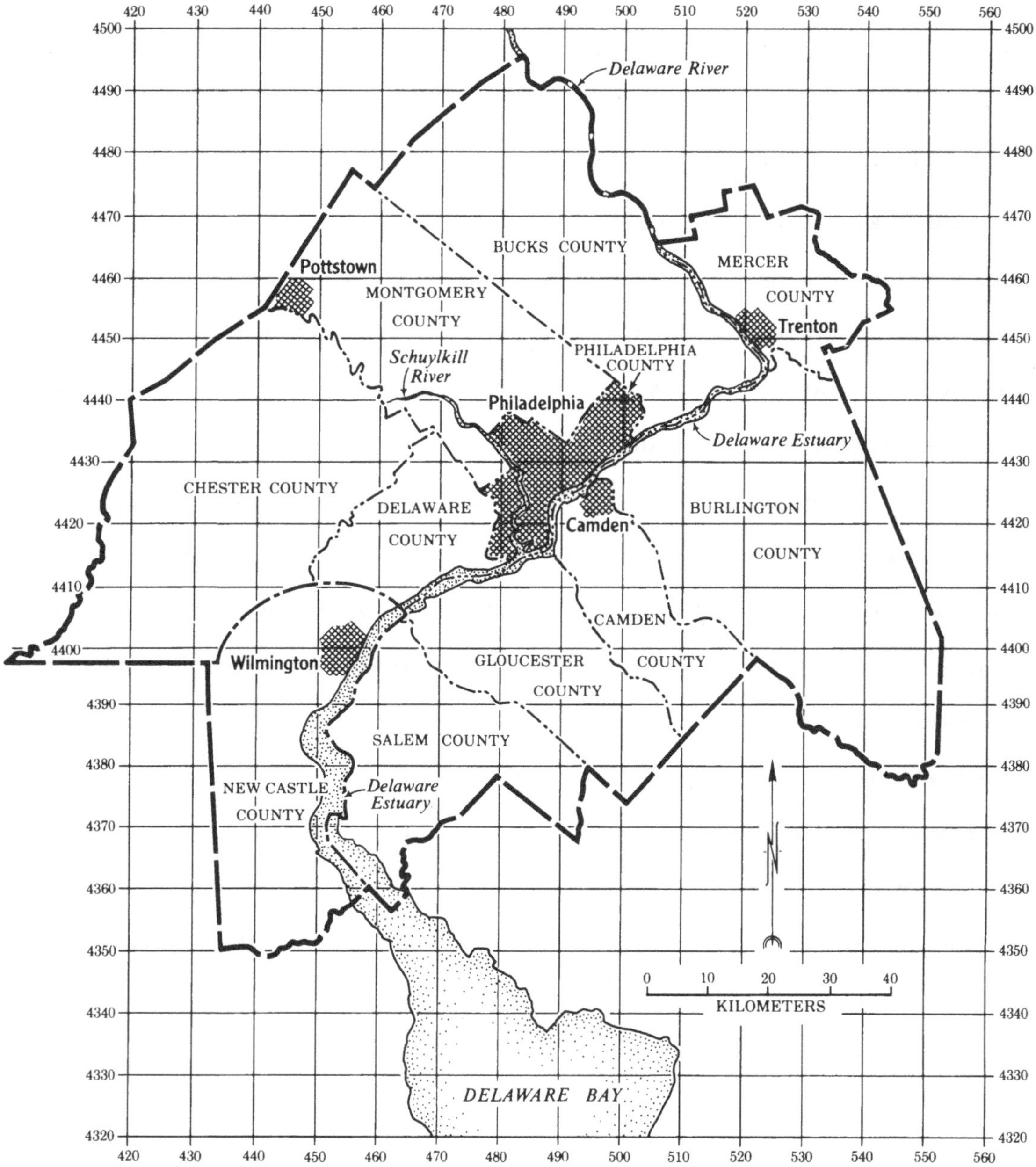

Region boundary
State line
County line

Note: The grid is in kilometers and is based on the Universal Transverse Mercator (UTM) Grid System.

Figure 1. Lower Delaware Valley Region.

and Salem Counties in New Jersey; and New Castle County in Delaware. Included in these 11 counties are 379 incorporated political jurisdictions--cities, towns, townships, boroughs, and divisions: 239 in Pennsylvania, 129 in New Jersey, and 11 in Delaware [20-22]. The major cities in the area are Philadelphia (coterminous with Philadelphia County) in Pennsylvania with a 1970 population of 1,948,609; Trenton, population 104,786 within the city bounds, and Camden, city population 102,551, in New Jersey; and Wilmington in Delaware with a 1970 population of 80,386 within the city limits [23-25]. The 11-county region comprises an area of roughly 4,700 square miles (12,200 square kilometers), less than half the size of Belgium or about the size of the State of Connecticut [26-29]. The region is about 80 miles (130 kilometers) across from West to East and roughly 90 miles (145 kilometers) from South to North. The rivers of interest in this region are: the reach of the Schuylkill from Pottstown, Pennsylvania, to the confluence with the Delaware at Philadelphia; the Delaware Estuary which runs approximately 85 miles from the head of tide at Trenton, New Jersey, to the head of Delaware Bay; a short reach of the Delaware River above Trenton; and roughly 30 tributaries of the Delaware and the Schuylkill within this part of the Delaware River Basin [30].

This region, with a 1970 population of approximately 5.6 million people is one of the most heavily industrialized areas in the United States. It contains 7 major oil refineries; 5 steel plants (1 employing open hearth furnaces; 1 with basic oxygen furnaces, and 3 having electric arcs) and many smaller steel-related operations; 15 large thermal electric power plants (and two smaller ones); a host of large and small chemical and petrochemical plants of various types; 13 pulp and paper or paper mills with more than 100 tons per day output, and numerous smaller ones; and many other types of industrial plants including cement and asphalt operations, assembly plants for the automobile and electronic industries, and food-processing operations.

In 1968, there were 17 municipal sewage treatment plants in the region with flows greater than one million gallons per day (mgd) discharging to the Delaware Estuary. (There were 7 plants with flows greater than 10 mgd.) In addition, 123 plants of various sizes discharged to the Schuylkill River and to the tributaries to the two major rivers. In 1970, there were 17 municipal incinerators in operation and numerous landfills and dumps in the region.

The Surface Waters of the Region

The flow of the Delaware River varies widely, from month to month and year to year. During the period 1951 through 1972, for example, the monthly mean river flow at Trenton varied from a low of about 1,550 cubic feet per second (cfs) in July of 1965 to a high of nearly 40,000 cfs in late spring of 1958. The mean annual flow for this period ranged from a low of about 4,700 cfs for 1965 to a high of about 18,000 cfs for 1952 [30]. The mean discharge at Trenton over the 60-year period from 1912 to 1972 is about 11,400 cfs.[3] The low-flow period, and hence the period during which the river is most sensitive to residuals discharges, is July-October.

The river flow at Trenton used as the basis for the aquatic

[3] Information provided by the Water Resources Division, U.S. Geological Survey, Reston, Virginia.

ecosystem model (discussed later on) corresponds to the mean flow for September 1970: 4,146 cfs (2.8 times as large as the 7-day, 10-year low flow at Trenton of 1,503 cfs and 1.3 times as large as the "design" flow of 3,303 cfs used by the Delaware River Basin Commission in their analyses).

The Schuylkill River is considerably smaller than the Delaware, with a mean discharge over the 47-year period from 1926 to 1972 of about 1,800 cfs at Pottstown, Pennsylvania, approximately where the Schuylkill enters the Lower Delaware Valley region. (The mean discharge for the Schuylkill at Philadelphia over a 34-year period was approximately 2,850 cfs.) The mean annual flow for the period 1951 through 1965 varied from a low of 843 cfs (1965) to a high of 3,059 cfs (1952). The monthly mean flow during this same period ranged from a low of 324 cfs in October 1963 to a high of 4,758 cfs in February 1951 [30]. The 7-day, 10-year low flow of the Schuylkill at Pottstown is about 240 cfs.[4]

The major tributaries in the Schuylkill-Delaware river basins are still smaller, ranging in mean annual flows from a high of about 430 cfs (Brandywine Creek at Wilmington, Delaware) to a low of roughly 4 cfs (Blackbird Creek in Blackbird, Delaware). The flows in many of these tributaries are so low in some months of the summer that return flows from wastewater treatment plants currently comprise a major portion of the low flows. Management alternatives that would reduce these low flows still further, such as regional sewage treatment schemes, therefore may not be desirable. Preliminary calculations indicate that the flows of at

---

[4]Information provided by the Water Resources Division, U.S. Geological Survey, Reston, Virginia.

least 20 tributaries in the region would be seriously affected by regional wastewater treatment schemes if compensating flows were not provided.

The Atmosphere and Climate of the Region

The atmospheric "resource" of the region does not lend itself to so simple a characterization as the flow relations for the watercourses. For the region as a whole, the seasonal prevailing wind pattern is roughly: winter and spring westerlies (and west-northwesterlies), summer southwesterlies, and autumn variability. In almost every month there are soutwesterly winds along the estuary stretch from roughly the Delaware (state) line to Camden, New Jersey. The net effect of these meteorological conditions, together with the spatial pattern of discharges of gaseous residuals, is a "mountain" of air pollution along the estuary from Wilmington to Trenton, with the highest concentrations over the Philadelphia-Camden area [31 (Table H-5)].

During the period 1936 to 1965, atmospheric stagnations[5] over Philadelphia occurred most often during the month of September with 16 cases recorded out of a total of 33; the next most frequent month was October with 8 cases [32]. Air pollution episodes are often associated with atmospheric stagnations, although the former are also dependent upon residuals generation and discharge rates during these adverse periods. During the two-year period 1957-1959, 22 air pollution episodes were recorded in the Philadelphia area: 4 in the spring; 2 in the summer; 8 in the fall; and 8 in the winter [33]. From information on the frequency of

---

[5]Atmospheric stagnations are defined here as a period of four or more days where the windspeed at anemometer level does not exceed 7.5 knots [32].

atmospheric stagnations alone, proportionally more episodes would have been expected in the fall and fewer in the winter. However, discharges of sulfur dioxide and particulates in the Philadelphia area are typically higher in the winter than during any other period of the year [31, Appendix H].

The ventilation factor (the product of the mean wind speed and the atmospheric mixing depth) is often used as a surrogate for short-term predictions of air pollution conditions. According to the mean monthly ventilation factor for Philadelphia computed from published data [34,35], the assimilative capacity of the air is highest in the summer, then spring, followed by fall, and is lowest during the winter. This is shown in Table 1. On average, the best environmental conditions for the dilution and dispersion of gaseous residuals in the Philadelphia area appear to be in June, the worst in January.

Other environmental data that aid in describing the monthly variation in the climate of the Philadelphia region--temperature, heating degree days, sea-level atmospheric pressure, and precipitation--are included in Table 2 [36]. These data will be referred to later in presenting the application to the Lower Delaware Valley region.

Discharges of Liquid and Gaseous Residuals

Residuals (waste loads) entering the Delaware Estuary originate from four sources: industries, municipalities, tributaries, and storm water runoff. Of the estimated 936,000 pounds per day of biochemical oxygen demand (BOD) that entered the estuary during the month of September 1970, approximately 41 percent derived from industries, 46 percent from municipal discharges, 5 percent from tributary loads, and 8 percent from storm water runoff. This is shown in Table 3a. Also indicated in

Table 1

Computed Mean Monthly Ventilation Factor
-- Philadelphia --

Month	Mean monthly wind speed[a] V (mph)	Monthly mean maximum atmospheric mixing depths[b] D (meters)	Computed mean monthly ventilation factor VD ($\times 10^{-3}$)
January	10.1	400	4.0 (min)
February	11.0	400	4.4
March	12.2	800	9.8
April	11.2	800	9.0
May	8.9	1,000	8.9
June	9.0	1,200	10.8 (max)
July	7.8	1,200	9.4
August	7.2	1,000	7.2
September	7.9	1,000	7.9
October	8.5	600	5.1
November	9.5	600	5.7
December	10.0	500	5.0
Annual	9.4[c]	792[c]	7.5[c]

[a]J. R. Mather, "Meteorology and Air Pollution in the Delaware Valley," C. W. Thornthwaite Associates (for the Regional Conference of Elected Officials, Inc.), Elmer, New Jersey, November 1967, Table 2, p. 25.

[b]G. C. Holzworth, "Estimates of Mean Maximum Mixing Depths in the Contiguous United States," Monthly Weather Review, Vol. 92, No. 5 (May 1964), pp. 235-242, Figures 2-13.

[c]Sum of column divided by 12.

Table 2. Climatic Date for Philadelphia

Month	Normal daily temperature, °F [a]			Normal total heating degree days[b] (base 65°F)	Normal see level pressure[c] (millibars)	Normal monthly precipitation[d] (inches)
	Maximum	Average	Minimum			
January	40	32	24	1,014	1,017	3.3
February	42	33	25	890	1,017	2.5
March	50	41	32	744	1,015	3.9
April	63	52	41	390	1,015	3.5
May	73	63	52	115	1,015	4.2
June	82	71	60	12	1,014	4.0
July	86	76	65	0	1,016	4.3
August	84	74	64	0	1,016	4.9
September	77	67	56	60	1,018	3.4
October	67	56	45	291	1,018	3.2
November	54	44	35	621	1,018	3.5
December	42	34	26	964	1,019	3.1
Annual: total or average	63[e]	54[e]	44[e]	5,101	1,017	43.8

[a]Based on a 94-year record; taken from U.S. Department of Commerce (ESSA), "Climatic Atlas of the United States," U.S. Government Printing Office, Washington, D.C., June 1968, pp. 1-24.

[b]Ibid., p. 36.

[c]Ibid., p. 80.

[d]For Southeast Pennsylvania. Ibid., pp. 45-48, 51, 52.

[e]Sum of column divided by twelve.

Table 3a

Estimated Residuals Loads to Delaware Estuary, September 1970[a]
(1,000 pounds per day and percent of total)[b]

	Biochemical Oxygen Demand, BOD_5	Nitrogen, N	Phosphorus, P
Industrial[c]	385 (41%)	117 (53%)	7.9 (15%)
Municipal[d]	425 (46%)	63.8 (29%)	27.0 (50%)
Tributaries[e]	50.3 (5%)	19.1 (9%)	13.0 (24%)
Storm water[f]	76.0 (8%)	20.5 (9%)	5.8 (11%)
Total (rounded)	936 (100%)	220 (100%)	53.7 (100%)

Table 3b

Spatial Distribution of September 1970 Residuals Loads[a]

6 reaches receive 83 percent of the total BOD load
6 reaches receive 79 percent of the total nitrogen (N) load
6 reaches receive 80 percent of the total phosphorus (P) load

See next page for footnotes.

Footnotes to Tables 3a and 3b:

[a]Aggregated from data presented in Table 3 of W. O. Spofford, Jr., C. S. Russell and R. A. Kelly, "Operational Problems in Large-Scale Residuals Management Models," in Edwin S. Mills, ed., Economic Analysis of Environmental Problems (New York: National Bureau of Economic Research, 1975), pp. 210-211.

[b]Organic material is reported in pounds of 5-day biochemical oxygen demand (BOD_5) per day; nitrogen, in pounds of nitrogen (N) per day; and phosphorus, in pounds of phosphorus (P) per day.

[c]Industrial loads were estimated from unpublished data supplied by the Delaware River Basin Commission, Trenton, New Jersey.

[d]Based on 1968 sewage treatment plant discharges in Delaware River Basin Commission's Final Progress Report: Delaware Estuary and Bay Water Quality Sampling and Mathematical Modeling Project, May 1970, Figure 12. Nitrogen and phosphorus loads were estimated from the BOD loads according to data presented in G. A. Rohlich and P. D. Uttormark, "Wastewater Treatment and Eutrophication," in Nutrients and Eutrophication, Special Symposium Volume 1, American Society of Limnology and Oceanography, Inc. (1972), pp. 231-245.

[e]Tributary loads are based on the low-flow season averaged over a three-year period.

[f]BOD loads are based on 1964 storm water overflow, in Robert V. Thomann, Systems Analysis and Water Quality Management (New York: Environmental Science Services Division of Environmental Research and Applications, Inc., 1972). Nitrogen and phosphorus loads have been estimated from BOD loads according to data given in U.S. Environmental Protection Agency, EPA Storm Water Management Model, U.S. Government Printing Office, Washington, D.C., 1971, p. 180.

this table are the sources of the nitrogen and phosphorus loads entering the estuary. Industry was clearly the largest contributor of nitrogen discharges to the estuary, whereas municipal effluents contained the largest amount of phosphorus [37-40].

Residuals discharges are distributed rather unevenly along the estuary. Six reaches (to be defined later) out of a total of twenty-two covering the 85 mile-long estuary receive 83 percent of the BOD load, 79 percent of the nitrogen load, and 80 percent of the phosphorus load (Table 3b).

The major point source dischargers of sulfur dioxide and particulates in the region are the petroleum refineries, steel mills, and power plants. Collectively, their sulfur dioxide discharges amount to about 1,760 tons per day, or roughly 60 percent of the total (2,964 tons per day) discharged to the region (see Table 4). Their collective particulate discharges are relatively high too, being 250 tons per day or roughly 42 percent of the total for the region. The power plants in the region clearly represent the largest single source type for both sulfur dioxide and particulates.

Area sources of sulfur dioxide and particulates account for about 25 percent and 35 percent of the totals for the region, respectively. However, these sources typically discharge close to the ground and hence contribute proportionally more to ground level ambient concentrations than their discharges would indicate.

The Environmental Quality Problems of the Region

The environmental quality problems of the region involve all three residuals-receiving media: atmosphere, water bodies, and land. For

Table 4

Estimated Sulfur Dioxide and Particulate Discharges in the Lower Delaware Valley Region

(annual average in tons per day)

Discharger Category	Sulfur dioxide[a]	Particulates[a]
Point sources (1,031 stacks):		
Petroleum refineries	410	66
Steel mills	19	58
Power plants	1,332	126
Other point sources	439	133
Subtotal	2,200	383
Area sources (240 areas):		
Home heating[b]	214	25
Other area sources	550	191
Subtotal	764	216
Total, all sources	2,964	599

[a] Based on EPA's 1970 inventory of gaseous emissions for the Metropolitan Philadelphia Interstate Air Quality Control Region, supplied by the Division of Applied Technology, Office of Air Programs, U.S. Environmental Protection Agency, Durham, North Carolina.

[b] Estimated from the number of housing units and the home heating fuel types contained in the 1970 Bureau of the Census computer tapes.

example, in 1970 there were numerous open dumps, some burning, contributing to air pollution, to surface and groundwater pollution, and to a general degradation of the aesthetic quality of the region. In the past, the dissolved oxygen concentration of the Delaware Estuary below Philadelphia has frequently dipped below 1.0 mg/ℓ for extended periods of time during the summer months of July, August, and September. A profile of the dissolved oxygen concentration of the Delaware Estuary for July to September 1968 is shown in Figure 2 [37]. Finally, for the period 1967-1968, the maximum annual average concentration of suspended particulates for the region, measured at a station in Philadelphia, was about 150 micrograms per cubic meter (μgms/m^3). (This compares with the federal primary standard of 75 μgms/m^3.) For sulfur dioxide, the maximum annual average concentration measured at the same station during the same period was about 190 μgms/m^3. (This compares with the federal primary standard of 80 μgms/m^3.) The distribution of sulfur dioxide over the region, based on a modeling study done by the U.S. Environmental Protection Agency, is shown in Figure 3 [41]. The highest predicted mean annual concentration of sulfur dioxide amounted to 244 μgms/m^3 in Philadelphia, somewhat greater than the maximum measured value of 189 μgms/m^3 at another location upon which the model was calibrated.

IV. The Regional Residuals Management Model

General Characteristics

The Lower Delaware Valley regional residuals management model is designed to provide the minimum-cost way of producing an exogenously determined "bill of goods" at the individual industrial plants; of meeting

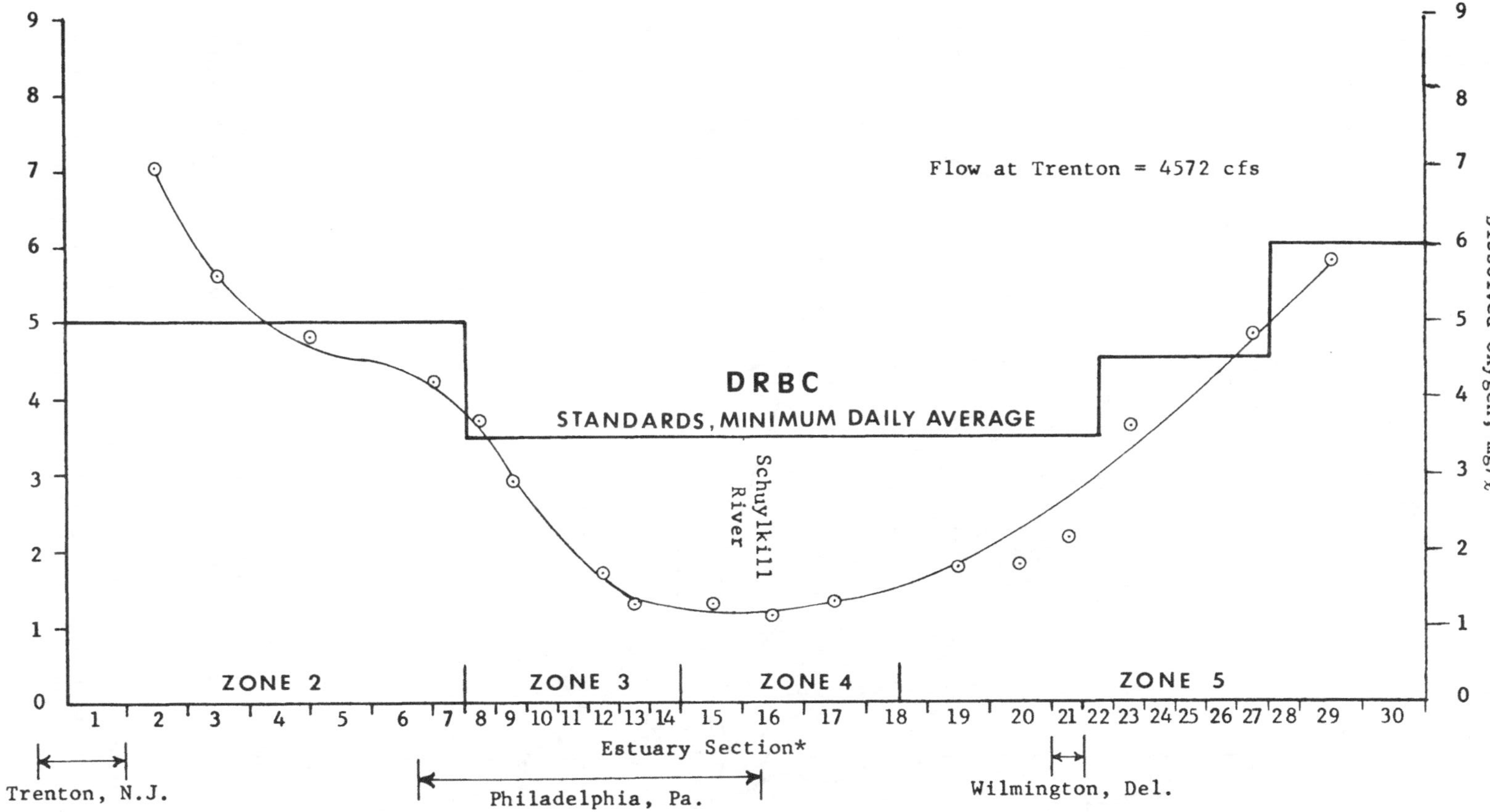

Figure 2. Delaware Estuary Dissolved Oxygen Profile: July to September 1968

Source: Delaware River Basin Commission, "Final Progress Report: Delaware Estuary and Bay Water Quality Sampling and Mathematical Modeling Project," May 1970, Figure 12.

Notes: *Delaware River Basin Commission sections. Note that these section designations differ from the ones in the RFF Study reported here.

——— predicted by the DRBC's Delaware Estuary Model.

⊙ mean of measurements for the 3-month period (July, August, September 1968)

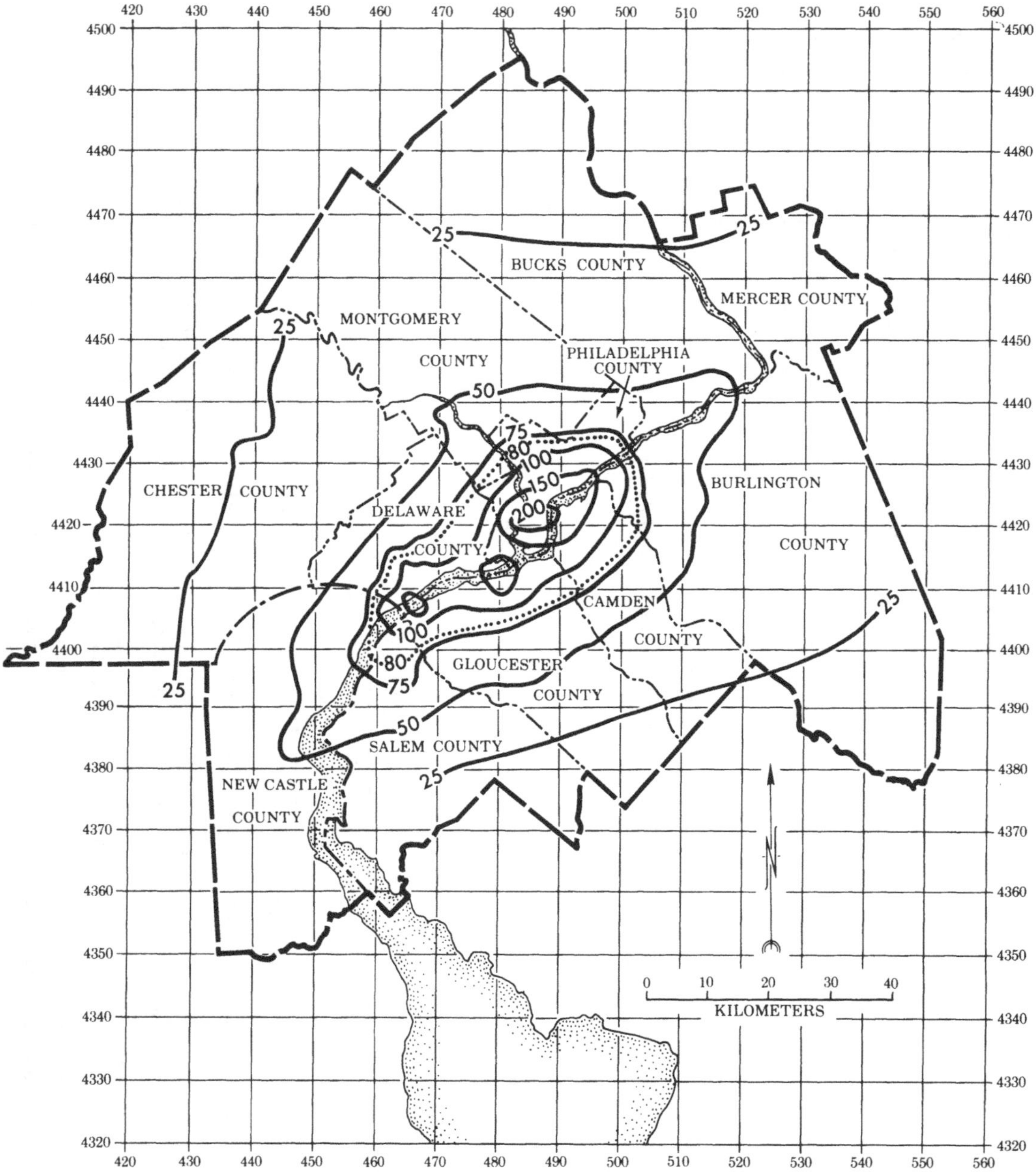

—25— Sulfur dioxide concentrations
▬ ▬ Region boundary
-·—·- State line
—··—··— County line

Notes:

1. Highest value predicted = 244 $\mu g/m^3$
2. Air quality standards:
   Primary = 80 $\mu g/m^3$ (dotted line)
   Secondary = 60 $\mu g/m^3$
3. The grid is in kilometers and is based on the Universal Transverse Mercator (UTM) Grid System.

Figure 3. Base Year Sulfur Dioxide Air Quality, 1967-68 ($\mu gms/m^3$)

Source: EPA, "Application of Implementation Planning Program (IPP) Modeling Analysis: Metropolitan Philadelphia Interstate AQCR," Air Quality Management Branch, EPA, Durham, N.C., February 1972, Figure 4.

electricity requirements and home and commercial space heating requirements for the region; and of handling, treating, and disposing of specified quantities of municipal liquid and solid residuals, subject to constraints on:

(a) the distribution of environmental quality (water, air, and landfills) over geographic units;

(b) the distribution of consumer costs (electricity, heating fuel, sewage disposal, solid waste disposal, regional in-stream aeration) over geographic units.

Credit is given in the model for the sale of newsprint and corrugated board produced from paperstock (used newsprint and used corrugated) collected within the region as an alternative to solid waste disposal, as well as for the sale of by-products produced at the steel mills and petroleum refineries in connection with reduction in residuals discharges.[6]

The main features of the model are:

1. It is a partial-equilibrium, optimization model.

2. It reflects nontreatment alternatives available for reducing the amounts of residuals initially generated.

3. It deals with the three major forms of residuals (liquids, gases, and solids) and the three receiving environmental media (water bodies, atmosphere, and land) simultaneously. The model reflects the conservation of mass and energy for relevant residual components (e.g., sulfur, waste heat, and municipal solid wastes), accounting for material and energy flows as they are modified from one form to another in

---

[6]The by-products include elemental sulfur, ammonium sulfate, and tar.

production, consumption, and residuals modification--or "treatment"--activities. Carbon dioxide and water vapor are ignored.

4. It is capable of incorporating various kinds of models of the natural world, from the simplest linear transformations (for example, the steady-state Streeter-Phelps-type dissolved oxygen models) to complex, nonlinear simulation models of aquatic ecosystems.

5. It is a static economic model so that time is not considered in the residuals generation and discharge portion of the model. Capacity expansion and optimal timing are not explored, no new industries enter the region, and the population level and spatial distribution remain constant. Therefore, a dynamic analysis of the impact on the region of economic and population growth cannot be made, but a comparative static analysis can be.

6. The spatial distribution of production and consumption activities in the region is fixed, although the impacts of various distributions could be examined and compared.

7. The natural world models are deterministic and steady state. (The ecosystem model is structured as a nonsteady-state model, but we use the eventual steady-state results.)

8. A single season (spring, summer, fall, winter, annual, episode conditions, or whatever) is employed for the analysis, although the model could be operated for different seasons, and the resulting residuals management strategies for each season compared.

9. Interactions among residuals in the atmosphere are assumed not to occur, and decay rates, where applicable, are independent of the quantities of the residual present (as well as of the presence of other residuals).

Model Structure and Solution Method

The management model is shown schematically in Figure 4. There are three main parts of this model: a linear programming (LP) model of regional residuals generation and discharge (comprising both production and consumption activities); the environmental models; and an environmental evaluation section. A key output of the LP model is a vector of residuals discharges, identified by substance and location. This vector is input to the environmental models--the aquatic ecosystem model and the air dispersion model. The environmental models produce as output a vector of ambient environmental quality levels (for example, SO_2 concentrations) at numerous designated points in the region. These concentrations are input to the final component, the environmental evaluation section. Here the ambient concentrations implied by a given solution of the LP model are compared with exogenously set environmental "standards." Marginal penalties, based on penalties for exceeding the environmental standards and on the environmental models, are computed in this component and returned to the LP model as trial prices, or effluent charges, on residuals discharges for the next solution of the LP model. An iterative procedure based on the gradient method of nonlinear programming and on the use of penalty functions for satisfying constraints has been developed to meet the ambient standards, within some specified tolerance, and to select a vector of residuals discharges that meets the constraints at least cost to the region [2,4,5].[7] It is this computational feature which allows the use

---

[7]The penalty function approach for eliminating constraints is a well-known technique in nonlinear programming. See, for example [42,43].

Figure 4

SCHEMATIC DIAGRAM OF THE LOWER DELAWARE VALLEY RESIDUALS–ENVIRONMENTAL QUALITY MANAGEMENT MODEL

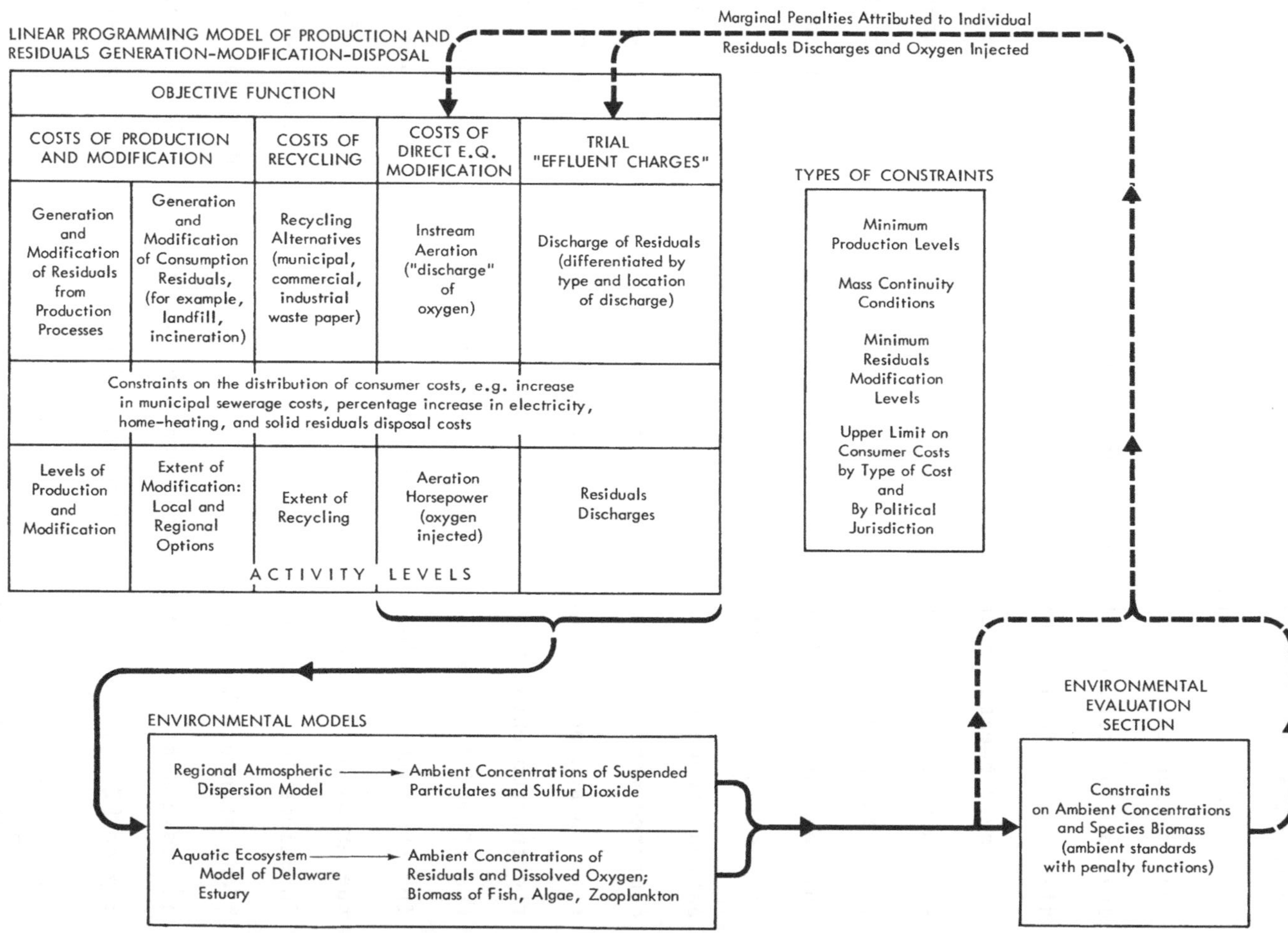

of nonlinear models of natural systems within an optimization framework. For a formal statement of the model and the solution method, see Appendix A.

Application of the Model to the Lower Delaware Valley

In the Lower Delaware Valley regional model, there are 297 sources of residuals--both point and nonpoint--that have been provided with management options for reducing their discharges.[8] The 183 point sources in the model include 124 industrial plants and 59 municipal waste handling and disposal activities--23 incinerators and 36 sewage treatment plants. The 114 area, or nonpoint, sources in the model include 57 home heating activities (one for each political jurisdiction) and 57 commercial heating activities (again, one for each jurisdiction). The other point and nonpoint sources distinghuished in the region that are not provided with residuals management options are incorporated as background discharges.

The 124 industrial plants, in turn, include the 7 oil refineries in the region, 5 steel plants, 17 power plants, and 23 other industrial dischargers to the estuary. A list of the industrial plants and municipal activities in the regional model is provided in Table B-1 of Appendix B.

The management options available to the various sources for reducing

---

[8]The major sources of information and data used in the construction of this model include: (i) EPA's 1970 inventory of gaseous emissions for the Metropolitan Philadelphia Interstate Air Quality Control Region for particulate and sulfur dioxide dischargers in the region; (ii) Delaware River Basin Commission for the municipal and industrial wastewater dischargers to the Delaware Estuary; (iii) Regional Conference of Elected Officials publications for municipal incinerators and for electricity districts and power plants; and (iv) the 1970 Bureau of the Census computer tapes for the consumer activities in the region--number of people and housing units by political jurisdictions, the home heating fuel characteristics, and sewerage systems. The base period for data employed in this model is 1968 to 1970.

their discharges range from production process and raw material input alternatives to residuals abatement devices for "end of pipe" treatment. This range is listed in Table B-2 of Appendix B for the various categories of sources in the model. We will return to a more complete discussion of the management options provided by category of discharger later on.

The sources of residuals that have been included in the regional model by no means represent an exhaustive list of all the residuals dischargers in the Lower Delaware Valley region. For example, the transportation and agricultural sectors, and land runoff, have not been included, except as "background" sources of residuals where appropriate. In addition, some of the smaller industrial plants in the region have not been provided with management options, but have also been included as "background" sources. The activities that are in the model with residuals management options available, however, include the important dischargers of sulfur dioxide and particulates to the atmosphere (as shown in Table 4) and the major dischargers of BOD, nitrogen, and phosphorus to the Delaware Estuary (Table 3a).

To represent the air quality of the region, annual average ground level concentrations of sulfur dioxide and suspended particulates were selected. This choice was based on human health considerations and on the availability of discharge data and an air dispersion model.[9] The water quality of the Delaware Estuary is represented by the dissolved oxygen

---

[9]The air dispersion model that we are using to predict sulfur dioxide and suspended particulates levels is taken from EPA's Implementation Planning Program (IPP) [44]. The discharge data are from EPA's inventory of gaseous emissions for the Metropolitan Philadelphia Interstate Air Quality Control Region.

level and by the biomass of algae and fish. The estuary is the only water body in the region that is described with a model. Originally, we had intended to incorporate in the regional model a water quality model of the Schuylkill River which was being developed by others, but that came along too late for us to use.[10] Consequently, dischargers up the Schuylkill are dealt with in the model with rather arbitrarily selected effluent standards (constraints). Unfortunately, this arrangement does not allow for liquid discharge tradeoffs in this part of the region. Dischargers to other tributaries are dealt with in the same manner as dischargers to the Schuylkill. The stretch of the Delaware River above Trenton currently is of fairly good quality, and there is little prospect for major degradation in the near future. Consequently, we decided early on not to incorporate a water quality model of this part of the region.

Regional options incorporated in the model for improving environmental quality include in-stream aeration, regional sewage treatment, and recycling of used newsprint and corrugated containers.

To provide information on the geographic distribution of environmental quality and consumer costs throughout the region, the major portion of the Lower Delaware Valley has been divided into 57 political jurisdictions of roughly 100,000 people each (the smallest: 60,346 people--jurisdiction 57; the largest: 112,621 people--jurisdiction 40). To form these jurisdictions, some of the 379 cities, towns, boroughs, and townships that are in this region have been aggregated; others have been subdivided, and

---

[10]Teledyne Isotopes, Inc. of Westwood, New Jersey, was in the process of constructing a water quality model of the Schuylkill River for EPA when we concluded our model development. Time did not permit us to wait until the model was available.

still others omitted [45]. In all cases, however, the 57 political jurisdictions are made up of whole census tracts and are located entirely within the boundaries of individual counties and hence within states. The identification and locations of these political jurisdictions, grouped by county, are shown in Table 5. Also indicated in this table are the 1970 populations and numbers of housing units by county. An expanded table is provided in Appendix B (Table B-3).

For each political jurisdiction, levels of air quality (annual average ground level concentrations of suspended particulates and sulfur dioxide at a central location), landfill quality, and levels of increased consumer costs resulting from the protection of the environment (for electricity, home and commercial heating, sewage disposal and solid waste disposal) are available, and their allowable upper limits can be specified. The air quality receptor locations for the 57 political jurisdictions are presented in Table B-4 of Appendix B. The quality of the estuary is available by reach (to be discussed below) and is related to the 57 political jurisdictions only indirectly.

There are three principal reasons for providing the data and constraints on the distribution of the costs of improving regional environmental quality. First, for any ambient quality standard set being investigated, the otherwise unconstrained model solution will tell us something about how the costs implied by that set are distributed, at least in the first instance, over the region. Second, by using the distributional constraints, we can explore how much aggregate cost increases when we change that initial distribution of costs. Third, we hope that in a legislative

Table 5

Political Jurisdiction for the Lower Delaware Valley Model

Jurisdiction identification	Location		Population[a] (1970)	Housing units[a] (1970)
	County	State		
1 - 4	New Castle	Delaware	385,856	120,704
5 - 7	Chester	Pennsylvania	252,152	73,154
8 - 13	Delaware	Pennsylvania	600,035	184,440
14 - 33	Philadelphia	Pennsylvania	1,920,206	662,265
34 - 39	Montgomery	Pennsylvania	623,799	193,592
40 - 43	Bucks	Pennsylvania	415,056	121,710
44 - 46	Mercer	New Jersey	302,472	96,429
47 - 49	Burlington	New Jersey	275,993	81,676
50 - 54	Camden	New Jersey	449,457	140,755
55 - 56	Gloucester	New Jersey	172,681	51,708
57	Salem	New Jersey	60,346	19,598
Total for the 57 Political Jurisdictions:			5,458,053	1,746,031
Area not covered by Political Jurisdictions 1-57:[b]			110,031	27,573
Total for the 11-county Lower Delaware Valley region:			5,568,084	1,773,604

[a]Population and housing counts are based on information in the 1970 Bureau of the Census computer tapes.

[b]For the political vote-trading portion of the model not all the people in the 11-county Lower Delaware Valley are represented. The roughly 110,000 people residing in the drainage basins of the Susquehanna and Raritan Rivers, and in the portion of New Jersey that drains directly to the Atlantic Ocean, are not represented.

setting the model with distribution information will allow a more efficient search for "vote trades" and hence for consensus policy.[11]

Arrangement of Economic Activities in the Regional Model

The residuals generation and discharge portion of the Lower Delaware Valley model is composed of six separate LP models arranged as depicted in Table 6. The first column of this table indicates the module number (for identification purposes only). The next three colums indicate the sizes of the individual LP modules--number of rows, number of columns, and number of residuals discharges, respectively. Note that in total there are almost 8,000 variables (columns) in the model and over 3,000 constraining relationships (rows) and that almost 800 individual residuals discharges enter the environmental models (to be discussed below). The residuals include BOD, nitrogen, phosphorus, toxics, suspended solids, and heat discharges to the estuary; and sulfur dioxide and particulate discharges to the atmosphere. The fifth column describes the type, and indicates the number (in parentheses), of activities in the region that are included in the model with residuals management options. These activities are identified in Table B-1 of Appendix B.[12] Management options available to the activities in the model are set out in Table B-2 of Appendix B and are discussed below. The sixth, and final, column of Table 6 depicts the extra cost distributional constraints available in each model. Except for the costs of sewage disposal, there is one constraint for each type

---

[11] For a fuller discussion of this last point, see [3].

[12] The regional activities in Module 2 are not identified in Appendix B. The home and commercial heating activities in this module are defined by political jurisdiction (Tables 5 and B-3).

Table 6

Lower Delaware Valley Model: Residuals Generation and Discharge Modules

Module identification	Size of linear program			Description[a]	Percent extra cost constraints for the 57 political jurisdictions (except as noted)[b]
	Rows	Columns	Discharges		
MPSX 1	286	1,649	130	Petroleum refineries (7) Steel mills (5) Power plants (17)	 57 electricity
MPSX 2	741	1,482	114	Home heat (57) Commercial heat (57)	57 fuel 57 fuel[c]
MPSX 3	564	1,854	157	"Over 25 μgms/m^3" dischargers (75)[d]	
MPSX 4	468	570	180	Delaware Estuary sewage treatment plants (36)	46 sewage disposal ($ per household per year)[e]
MPSX 5	951	1,914	88	Paper plants (10) Municipal incinerators (23) Municipal solid residuals handling and disposal activities	 57 solid residuals disposal
MPSX 6	229	395	117[f]	Delaware Estuary industrial dischargers (23)[g] In-stream aeration (22)	 57 in-stream aeration (absolute extra cost per day)[h]
TOTAL	3,239	7,864	786		

See next page for footnotes.

Footnotes to Table 6:

[a]The numbers in parentheses indicate the number of plants or activities that are included in the module with residuals management options.

[b]The numbers in this column indicate the number of distributional constraints of a specified type that are incorporated in the model.

[c]The commercial heating requirements in this module are based on the differences between sulfur dioxide discharges from area sources in the IPP inventory of gaseous emissions and sulfur dioxide emissions from the home heating model. Commercial heating requirements for eight political jurisdictions are equal to zero (nos. 19, 28, 29, 31, 32, 33, 37, and 44) [47,48].

[d]Industrial plants whose gaseous discharges result in maximum annual average ground level concentrations equal to or greater than 25 $\mu gms/m^3$. To determine this group, all stacks (point sources) in the IPP inventory were considered except those in MPSX 1 above (petroleum refineries, steel mills, and thermal power plants). The maximum annual average ground level concentrations (of sulfur dioxide and suspended particulates) were computed for each stack. For all stacks at the same x-y location (i.e., same plant), the maximum ground level concentrations were added together. Those plants associated with maximum ground level concentrations--for either sulfur or suspended particulates--equal to or greater than 25 $\mu gms/m^3$ were placed in this category.

[e]Only 36 of the 57 political jurisdictions discharge all their sewage directly to the Delaware Estuary. Another 10 discharge part of their sewage to the estuary. The remaining 11 jurisdictions that do not discharge to the estuary at all include: nos. 5, 6, 7, 36, 39, 43, 45, 46, 53, 54, and 56. The extra costs reported represent the average increase per household per year for each political jurisdiction.

[f]Does not include the 22 oxygen "discharges" from the in-stream aeration option.

[g]Twelve of the Delaware Estuary industrial wastewater dischargers in MPSX 6 are also represented by the sulfur dioxide and/or particulate dischargers in MPSX 3, and the gaseous discharges of another are included in MPSX 5.

[h]The model currently reports the total regional absolute extra cost per day for in-stream aeration. This cost is then allocated equally among the 57 political jurisdictions. Any other distribution is also possible.

Note: For the major sources of information and data used in the construction of the regional model, see footnote 8 on page 23.

of extra cost for each of the 57 political jurisdictions into which we divided the Lower Delaware Valley for modeling purposes. There are only 46 extra costs for sewage disposal in the model because 11 jurisdictions do not discharge any sewage at all to the estuary. To provide the reader with a flavor of the contents of the residuals generation and discharge portion of the regional model, we shall describe each module briefly.

Module 1 -- Petroleum Refineries, Steel Mills, and Power Plants.

The first module includes models of 7 petroleum refineries, 5 steel mills, and 17 power plants. These 29 plants are identified in Table B-1 of Appendix B. The 7 petroleum refinery models incorporated in the regional model are based on an RFF study of this industry [13]. All seven models are condensed versions of the same large refinery model and represent typical gasoline refineries.[13] Each model, however, has been scaled to the production level of the plant that it is meant to represent. The locations of the seven plants in the regional model are based on EPA's inventory of gaseous emissions for the region. The management options reflected in the models for reducing, or modifying, residuals discharges are listed in Table B-2 of Appendix B.

The 5 integrated iron and steel mill models incorporated in the regional model are also based on an RFF study [14,15], and also represent condensations of an existing larger model. Each model has been scaled to the production level of the plant that it is meant to represent, and each has been modified to reflect the type, or types, of steel furnace(s) that

[13] The condensation process is described in [4] and involves repeated solution of the large model, with these solutions each being summarized in a vector for inclusion in the small model. The effect is to reduce row size by a factor of about 20.

plant employs--open hearth, BOF, and electric arc. As with the petroleum refineries, steel mill locations in the regional model are based on EPA's inventory of gaseous emissions. The residuals management options provided in the model for this industry are also summarized in Table B-2 of Appendix B.

The 17 power plant models incorporated in the Lower Delaware Valley regional model are based on information contained in EPA's Implementation Planning Program for air quality control [44]. Each of the 17 models has been scaled to the capacities and the load factors of the plants they are meant to represent. The residuals management options provided in the model for power plants, shown in some detail in Table B-2 of Appendix B, include: fuel switching (various sulfur contents of coal and residual fuel oil, and natural gas); installation of abatement devices for the removal of sulfur dioxide and suspended particulates; and installation of cooling towers for reducing, or eliminating, heat discharges to surface waters. The service areas of the major utilities in the region are taken from a report of the Regional Conference of Elected Officials [31]. These service areas, in conjunction with the 57 political jurisdictions discussed above, are used in apportioning the extra cost of electricity for the distributional constraints shown in Table 6.

For all 29 industrial models in the first module, it is assumed that the solid residuals--bottom ash, sludge, and slurries--can be disposed of either on-site or conveniently nearby without degrading either the surface or groundwaters. No links are provided in the model for the natural transport of these residuals between solid residuals disposal sites and the other environmental media.

Module 2 -- Home and Commercial Heating.

There are 57 home heating activities in the regional model, one for each political jurisdiction. The home heating model is based on household characteristics and fuel usage data collected by the U.S. Bureau of the Census during 1970. These data were subsequently aggregated for our purposes to the 57 political jurisdictions. All households in the region, including those in multiunit structures, are considered in the model. The types of fuels considered for home heating options include coal (1.5 percent sulfur content, 9.0 percent ash); two grades of distillate fuel oil: 0.25 and 0.55 percent sulfur; five grades of residual fuel oil: 0.5, 1.0, 1.5, 2.0, and 2.5 percent sulfur (for the apartment buildings only); and natural gas (see Table B-2 of Appendix B). The heat requirements are based on the average annual heating degree days for Philadelphia (5,101) and information on the average home heating requirements in the United States presented in a Public Health Service publication [46]. In this model, it is possible to convert the entire region to natural gas for home heating purposes.[14] In allowing this, we assume that a gas-pipe network already exists in the region and that the costs of tapping into this system are minimal. Costs of new furnaces, where required, are included in the computations.[15]

---

[14]We realize that this is unrealistic in the post-1973 world. We justify the retention of this option by observing that it almost certainly provides a lower bound on the cost of meeting the federal primary air quality standards. And, as we shall see below, even under this generous allowance, it appears to be difficult to meet those standards over parts of Philadelphia.

[15]A more detailed description of the home heating module is contained in an unpublished RFF internal report [47].

There are also 57 commercial heating activities in the regional model. Unlike the home heating activities, however, for which we had an independent source of information on fuel usage (U.S. Bureau of the Census), the commercial heating requirements could only be calculated as a remainder, using EPA's inventory of gaseous emissions and the previously calculated home heating emissions in the base case. As a result of this analysis, eight political jurisdictions have commercial heating requirements equal to zero. In addition, we assumed that in the absence of environmental quality controls, commercial sources would be using residual fuel oil with a sulfur content of 2.5 percent. Fuel alternatives provided for this activity are shown in Table B-2 of Appendix B. As with the home heating model, it is assumed that all commercial space heating can be converted to natural gas.[16]

Module 3 -- The over 25 $\mu gms/m^3$ Dischargers

The selection of industrial activities for inclusion in this category requires an explanation. EPA's inventory of gaseous emissions for this region contains 1,031 point sources (stacks) and 276 area sources. We felt that this was far too many gaseous emission sources to be incorporated within the regional model with individual management options. To reduce the number of sources to be dealt with, we did two things. First, we aggregated all individual stacks at the same plant location. The aggregation procedure is discussed in more detail later on. For the 29 industrial plants in the first module, for example, we were able to reduce the 285 stacks in EPA's inventory to only 29.

---

[16]A more detailed description of the commercial heating module is contained in an unpublished RFF internal report [48].

Second, we selected for inclusion in the regional model only the important dischargers in the region using contribution to maximum annual average ground level concentrations as our criterion. We arbitrarily selected an ambient concentration of 25 $\mu gms/m^3$ for either sulfur dioxide or suspended particulates as the cut-off point; thus the name of the module. To determine the plants in this group, all stacks (point sources) in EPA's IPP inventory were considered except those in the first module (petroleum refineries, steel mills, and thermal power plants). The maximum annual average ground level concentrations (of sulfur dioxide and suspended particulates) were computed for each stack. For all stacks at the same plant, the maximum ground level concentrations were added together. Those plants associated with maximum ground level concentrations--for either sulfur dioxide or suspended particulates--equal to or greater than 25 $\mu gms/m^3$ were placed in this category. A total of 75 such plants were found. These plants are identified in Table B-1 of Appendix B. The residuals management model for each plant is based on information contained in EPA's IPP for air quality control [44]. The development of these models, including the adaption for use in the regional model, is described in an unpublished paper [49]. The management options available for reducing discharges of sulfur dioxide and particulates are shown in Table 5-2 of EPA's report [44]. (See also Table B-2 of Appendix B). Liquid residuals discharges for these plants are not included in this module. For some of these plants, the liquid residuals are dealt with in modules 5 and 6.[17]

---

[17]Where a plant's options include such processes as wet stack gas scrubbing or sludge incineration, implying direct trades between receiving media, the necessary entries occur in a single LP module so that the optimization routine at each iteration will reflect the proper cost considerations.

Module 4 -- Delaware Estuary Sewage Treatment Plants.

This module contains 36 municipal sewage treatment plants that discharge liquid residuals to the Delaware Estuary (see Table B-1 of Appendix B). Information on the locations and sizes of these plants, and on the influent and effluent loadings, was furnished, for the most part, by the Delaware River Basin Commission (DRBC). Management alternatives provided in the model for reducing and/or modifying residuals discharges are shown in Table B-2 of Appendix B. Cost functions for these alternatives were developed from information available in the literature [50].

There are two versions of this module: one for the 36 individual municipal sewage treatment plants, and one that incorporates 2 regional sewage treatment plants--1 on the Pennsylvania side and 1 on the New Jersey side of the estuary. These 2 regional plants replace 7 individual plants. See Table 7 for details. For a given run of the regional management model, one or the other but not both, municipal sewage treatment plant modules can be employed. In this way, the implications of a regional sewage treatment scheme can be compared with that of the individual municipal plants. The regional treatment plant alternative is based on research done at the Johns Hopkins University [51].

Module 5 -- Paper Plants and Municipal Solid Residuals

This module comprises the regional solid residuals handling and disposal activities, including options for recycling used newsprint and corrugated containers, and some of the paper plants in the region.[18] We

---

[18]The development of this module is based on an earlier conceptual study done at RFF on solid residuals management and waste paper recycling [19].

Table 7

The Regional Sewage Treatment Plant Alternative--Module 4

Sewage treatment plant	Location			Delaware Estuary reach number[a]	
	City	County	State	RFF	DRBC
Regional sewage treatment plant No. 1 (replaces):	Ridley	Delaware	PA	13	17
1. Philadelphia South East STP	Phila-delphia	Phila-delphia	PA	10	14
2. Philadelphia South West STP	Phila-delphia	Phila-delphia	PA	12	16
3. Central Delaware Sewerage Authority STP	Ridley	Delaware	PA	13	17
4. Chester STP	Chester	Delaware	PA	14	18
Regional sewage treatment plant No. 2 (replaces):	Pennsville Twp	Salem	NJ	18	23
1. Penns Grove STP	Penns Grove	Salem	NJ	17	21
2. Upper Penns Neck STP	Upper Penns Neck Twp	Salem	NJ	18	23
3. Pennsville STP	Pennsville Twp	Salem	NJ	18	23

[a]RFF -- Resources for the Future estuary reach numbering scheme.
DRBC -- Delaware River Basin Commission reach numbering scheme.

distinguish between two types of paper plant: one that uses virgin pulp as the raw material input and one that uses paperstock (recycled paper). One paper plant of the first type (Scott Paper Company) and 14 of the second type are included in the module. These 15 plants are identified in Table B-1 of Appendix B.

We assume that all the used newsprint and corrugated containers input to these 14 plants is generated within the region. Because of differences in the costs of collecting used paper and board, we distinguish between two sources of corrugated board--industrial and wholesale/retail--and between two sources of newsprint--low density residential and high density residential.[19] Twelve of the paper plants in this module that use paperstock as the raw material input are constrained to meet certain production levels. These plants require as input a mixture of newsprint and corrugated, some with fixed ratios, others with more flexibility. Consequently, a certain amount recycling of both newsprint and corrugated is required by the model, regardless of costs.

In addition to these 12 plants, 2 new plants that use recycled fiber as input can be built--a newsprint plant and a linerboard plant. The newsprint plant requires used newsprint as input, the linerboard plant requires used corrugated. These plants are constructed only if the total costs to the region of paper production, net of credit for the sale of paper ($11.15 per ton for linerboard and $16.00 per ton for newsprint), and of solid residuals handling and disposal are reduced by this alternative. In the

---

[19]Low density residential is arbitrarily defined as four or fewer housing units per structure, and high density residential as more than four units per structure.

model, these 2 plants must buy paperstock from the more expensive sources. The less expensive sources are used up first to satisfy the requirements of the 12 other plants. In Table 8, we indicate the amounts of used newsprint and corrugated containers that are assumed to be available in the region, the amount required by the 12 existing plants that must meet a given production level, and the remaining paperstock available as input to the 2 new plants that can be built if it is economically efficient (on a regional basis) to do so.

For the municipal solid residuals management portion of the module, various options are available including transport to landfills within the region; incineration of municipal solid residuals and the disposal of the incinerator residue in landfills; railhaul of the solid residuals out of the region; and the separation and collection of newsprint for recycling purposes. The used corrugated from commercial sources can be transported to landfills within the region, disposed of outside the region using the railhaul alternative, or recycled.

The landfills in the region can be of three qualities with different costs in the model for achieving each. Landfill qualities in each of the political jurisdictions can be controlled independently by the user of the regional model. In some cases, political jurisdictions dispose of their solid residuals in other jurisdictions (for example, Philadelphia disposes part of its solid residuals in Delaware County), although, in general, disposal is within the jurisdiction in which the residuals are generated.

There are 23 municipal incinerators in this module, ranging in existing capacity from 25 to 600 tons per day. Only 13 of the incinerators, with an aggregate capacity of 4,890 tons per day input, actually exist

Table 8

Assumed Availability of Used Newsprint and Corrugated Containers in the Lower Delaware Valley Region

(tons/day)

Source/Use	Corrugated	Newsprint
Total available in the region:		
1. Industrial sources	414	
2. Wholesale/retail sources	1,128	
3. High density residential[a]		112
4. Low density residential[a]		732
	1,542	844
Required by the 12 plants in region using paperstock[b]	1,235	141
Available for new plants:		
1. Linerboard	307	
2. Newsprint		703

[a]Low density residential is arbitrarily defined as 4 or fewer housing units per structure, and high density residential as more than 4 units per structure.

[b]Although various mixes of newsprint and corrugated can be used as input to these 12 plants, these use levels were observed in 15 of the first 16 computer runs of the model (to be discussed later on).

in the region [31]. The model allows for 10 new incinerators to be built, if it is economically feasible to do so, in order to increase the incineration capacity within the region. The total aggregate capacity for all 23 incinerators is 9,340 tons per day. The management options provided in the model for reducing discharges of particulates to the air and suspended solids to the water from municipal incinerators are shown in Table B-2 of Appendix B.

Module 6 -- Delaware Estuary Industrial Dischargers

The 23 industrial plants in this module discharge their liquid residuals to the Delaware Estuary. These industries are identified in Table B-1 of Appendix B. They are grouped together because they constitute one of the major sources of industrial discharges to the estuary according to DRBC records. The other major industrial source are the refineries, the steel mills, and the power plants which are included in Module 1. Information on the residuals loads and locations of these plants were provided by the DRBC. The management options provided in the regional model for reducing, or modifying, liquid residuals discharges from these sources are similar to those provided for the municipal sewage treatment plants (Module 4) and are based on cost and removal efficiency estimates from the literature. These management alternatives are listed in Table B-2 of Appendix B.

The in-stream aeration option is also incorporated in this module by means of 22 in-stream aeration activities, one for each reach of the estuary. The maximum total capacity of the aerators per reach is 1,000 horsepower, except for reaches 17 and 18, both of which have a maximum total capacity of 1,500 horsepower. This module is based on the work of

Whipple and his colleagues at the Water Resources Research Institute, Rutgers University [52,53,54]. More detail on this option is contained in an unpublished RFF report [55].

The Delaware Estuary Model

The regional residuals management model incorporates a nonlinear ecosystem model of the Delaware Estuary [6,7,9]. For this purpose, the Delaware Estuary has been divided into 22 reaches which are shown in Figure 5. The ecosystem model is based on a trophic level approach. The components of the ecosystem are grouped in classes, or "compartments," according to their function, and each class is represented in the model by an endogenous, or state, variable.

Inputs of liquid residuals discharges to the ecosystem model include: organics (BOD), nitrogen, phosphorus, toxics (phenols), suspended solids, and heat. Outputs are expressed in terms of ambient concentrations of algae, bacteria, zooplankton, fish, dissolved oxygen, BOD, nitrogen, phosphorus, toxics, suspended solids, and temperature. Three of these outputs--algae, fish, and oxygen--are constrained. These constraint levels are set exogenously to the management model and represent both a major driving force in the solution runs (when they are "binding") and one of the principal policy issues the model is designed to study. Thus, the cost of meeting alternative constraint sets on the levels of dissolved oxygen and of the biomass of fish and algae can be viewed as one of the important capabilities of the regional model.

A summary of the estuary model indicating relevant inputs and outputs is presented in Table 9, and the material flows among components (or "compartments") of the ecosystem are depicted in Figure 6. This

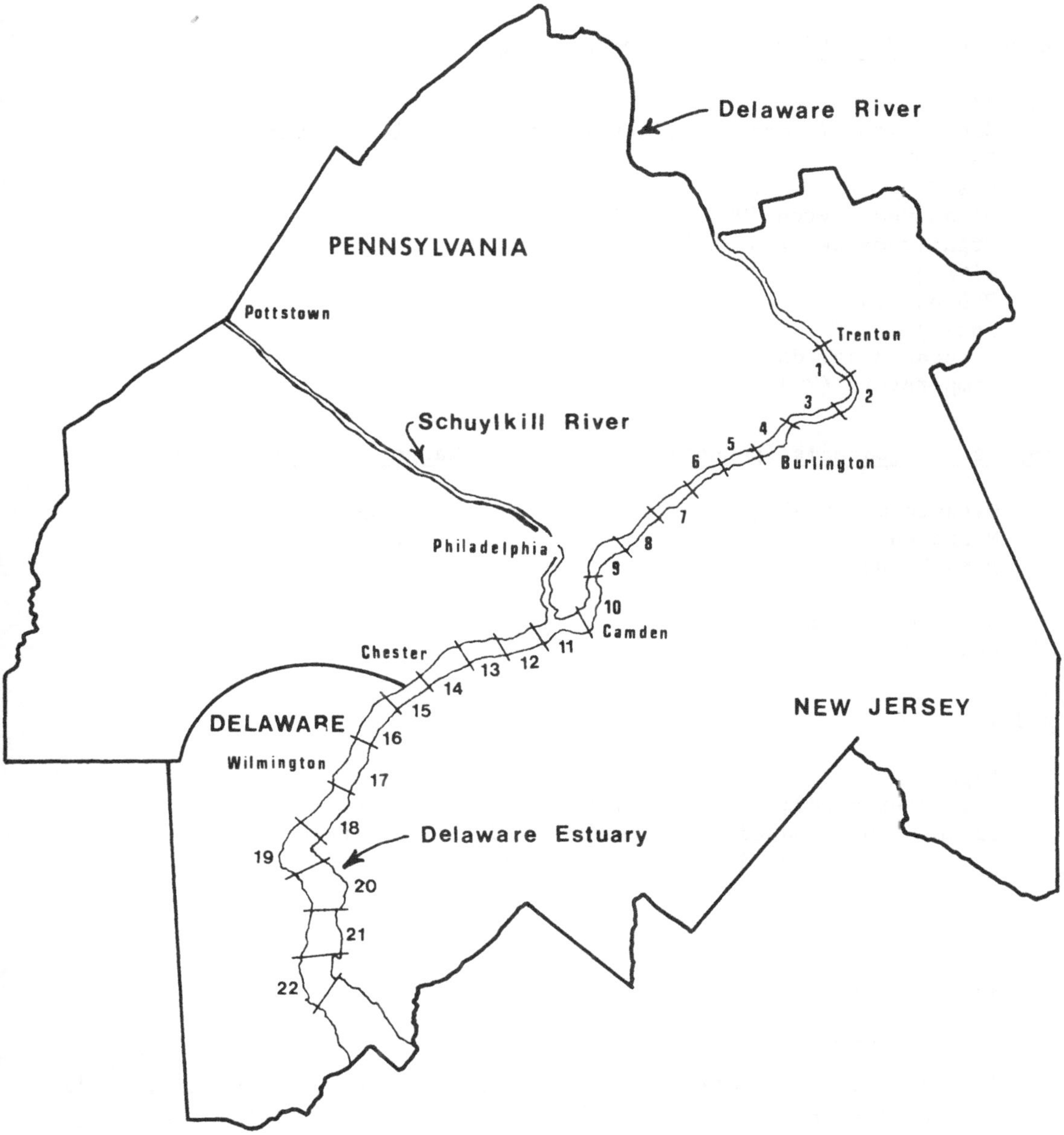

Figure 5. Reach locations for Delaware Estuary Model

Source: R. A. Kelly, "Conceptual Ecological Model of the Delaware Estuary," in B. C. Patten, ed., Systems Analysis and Simulation in Ecology, Vol. IV (New York: Academic Press, forthcoming).

Table 9

Delaware Estuary Ecosystem Model[a]

Endogenous Variables (compartments -- mg/ℓ)

Algae
Zooplankton (herbivores, detritivores, and bacterivores)
Bacteria
Fish
Dissolved oxygen (DO)
Organic matter (as BOD)
Nitrogen
Phosphorus
Toxics
Suspended solids
Temperature (^{o}C)

Inputs of Residuals (pounds/day)

Organic material (as BOD)
Nitrogen
Phosphorus
Toxics
Suspended solids
Heat (Btu)

Target Outputs[b] (mg/ℓ)

Algae
Fish
Dissolved oxygen

Model

Type: materials balance-trophic level
Characteristics: deterministic, nonsteady-state
Calibration: based on September 1970 flow at Trenton, New Jersey, of 4,146 cfs.

Reaches

Number: 22

---

[a]For details, see: (1) Robert A. Kelly, "Conceptual Ecological Model of the Delaware Estuary," to be published in Bernard C. Patten, ed., Systems Analysis and Simulation in Ecology. Vol. IV (New York: Academic Press, forthcoming); and (2) Robert A. Kelly and Walter O. Spofford, Jr., "Application of an Ecosystem Model to Water Quality Management: The Delaware Estuary," to be published in Charles A. S. Hall and John W. Day, Jr., eds., Models as Ecological Tools: Theory and Case Histories (New York: Wiley-Interscience, Inc., forthcoming).

[b]Management model operated for relevant minimum, or maximum, allowable ambient concentrations ("standards").

Notation: N = nitrogen; P = phosphorus; A = algae; B = bacteria; H = zooplankton; F = fish; L = organic matter (as BOD); 0 = dissolved oxygen

Note: The three remaining endogenous variables--heat (temperature), toxics, and suspended solids--are assumed to affect the rates of material transfers among the ecosystem components, or "compartments"

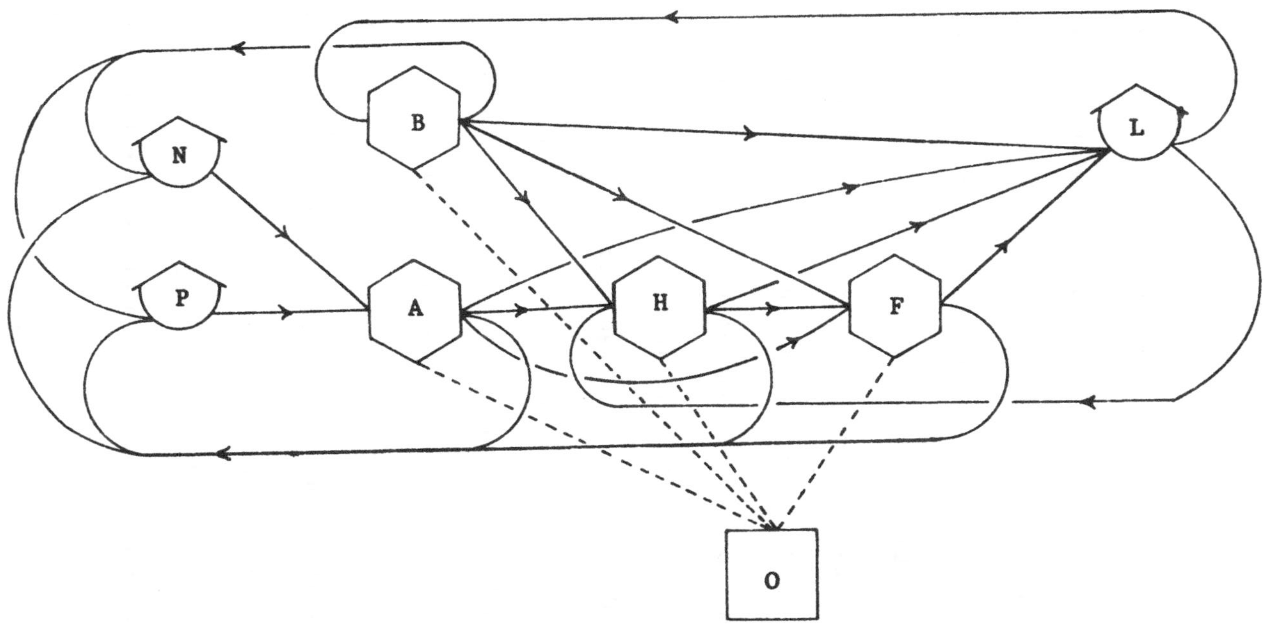

Figure 6. Diagram of materials flows among compartments within a single reach: Delaware Estuary Ecosystem Model

Source: Walter O. Spofford, Jr., Clifford S. Russell, and Robert A. Kelly, "Operational Problems in Large-Scale Residuals Management Models," in Edwin S. Mills, ed., <u>Economic Analysis of Environmental Problems</u> (New York: National Bureau of Economic Research, 1975).

model has been calibrated using September 1970 estuary flow and residuals discharge data.

The Regional Air Quality Model

In addition to the Delaware Estuary model, the regional management model includes two 57 x 240 (57 receptor locations, one for each political jurisdiction, and 240 dischargers) air dispersion matrices--one for sulfur dioxide and one for suspended particulates. These matrices relate annual average ambient ground level concentrations to discharges of sulfur dioxide and particulates. The matrices are based on output of the air dispersion model contained in EPA's Implementation Planning Program [44]. A summary of the air dispersion model, indicating relevant inputs and outputs, is given in Table 10.

The air dispersion model was calibrated using equations developed in an unpublished EPA study of the Philadelphia Metropolitan Air Quality Control Region [41].[20] The following atmospheric data were employed in the analysis of the air quality of the region:

1. Annual joint probability distribution for wind speed, wind direction, and atmospheric stability[21]

   Philadelphia -- 1970 data.

---

[20]The following calibration equations were used with the air dispersion model:

For sulfur dioxide: $y = 0.0 + 0.416x$

For suspended particulates: $y = 35.0 + 0.532x$

where y = calibrated ambient concentrations ($\mu gms/m^3$)
x = ambient concentrations predicted by uncalibrated IPP air dispersion model ($\mu gms/m^3$)

[21]Provided by the Applied Technology Division, Office of Air Programs, U.S. Environmental Protection Agency, Durham, North Carolina.

Table 10

Atmospheric Dispersion Model[a]

<u>Endogenous Variables</u> (ground level, annual average ambient concentrations-- μ gms/m^3)

Sulfur dioxide
Suspended particulates

<u>Inputs of Residuals</u> (tons/day)

Sulfur dioxide
Particulate matter

<u>Target Outputs</u>[b] (μgms/m^3)

Sulfur dioxide
Suspended particulates

<u>Model</u>

Type: Gaussian plume dispersion model from EPA's Air Quality Implementation Planning Program (IPP)

Characteristics: deterministic, steady-state

Input requirements:

1. <u>Sources</u>
   (a) x-y coordinates of each stack
   (b) discharge rates for each source
   (c) physical stack height
   (d) stack diameter
   (e) stack exit temperature
   (f) stack exit velocity
2. <u>Receptors</u>

   x-y coordinates of each receptor location
3. <u>Atmosphere</u>
   (a) annual joint probability distribution for wind speed (6 classes), wind direction (16 directions), and atmospheric stability (5 classes). The result is 480 discrete meteorological situations
   (b) mean annual temperature and pressure
   (c) mean annual afternoon atmospheric mixing depth

---

[a]For details, see TRW, Inc., "Air Quality Implementation Planning Program," U.S. Environmental Protection Agency, Vols. I & II, November 1970 (also available from National Technical Information Service, Springfield, VA 22161, as nos. PB-198 299 and PB-198 300, respectively).

[b]Management model operated for relevant maximum allowable ambient concentrations ("standards").

2. Mean annual temperature and pressure (see Table 2)

   Temperature[22] -- 68^oF (20^oC)

   Pressure -- 1,017 millibars (30.03 inches of mercury)

3. Mean annual maximum afternoon atmospheric mixing depth (see Table 1 and [35])

   1,000 meters[23]

The discharge dimension of the two air dispersion matrices requires an explanation. As we noted above, there are 240 inputs of gaseous residuals to the air dispersion matrices--183 from point sources and 57 from area sources.[24] In the residuals generation and discharge portion of the regional model, there are 114 area sources--57 home heating activities and 57 commercial heating activities, all located in Module 2 (see Table 6). For each political jurisdiction, the home and commercial heating discharges are added together prior to entering the air dispersion model (that is, prior to multiplication by the air dispersion matrices). We assume for air dispersion modeling purposes that the discharge heights above ground level for these two sources are the same.

---

[22]The mean temperature used in the model (68^oF) is roughly that of September. The mean annual temperature is actually somewhat lower than this (54^oF). See Table 2.

[23]The mean maximum afternoon atmospheric mixing depth used in the model (1,000 meters) is actually that of the months of May, August, and September. In June and July, the mean mixing depth is greater than 1,000 meters; from October to April, it is less than 1,000 meters (see Table 1). The effect of decreasing the mixing depth around the 1,000 meter level is to increase the concentrations predicted by the model.

[24]The 183 point sources derives from the 196 point source activities indicated in Table 6, less the 12 activities in Module 6 with gaseous discharges in Module 3 and the one activity in Module 6 with gaseous discharges in Module 5.

The point sources in EPA's inventory of gaseous emissions comprise a number of groups of stacks clustered at the same plant location. To keep the two air dispersion matrices as small, and thus as manageable, as possible, we decided that multiple stacks at the same location would be treated in the model as a single stack. Thus, only one column is required in each dispersion matrix for each plant in the regional model that is provided with residuals management options. To evaluate the dispersion coefficients in these columns, two methods were employed. For most of the plants in the region, a single stack was selected to represent the group of stacks. The coefficients in the air dispersion matrices associated with these plants were based on the output of EPA's air dispersion model using as input a single stack per plant. For two plants in the region, relative contributions to ground level concentrations were based directly on the output of EPA's air dispersion model using as input the group of stacks associated with each plant.[25] The first method of evaluating these coefficients is substantially less time consuming and less expensive than the second. However, the second method results in a more accurate set of air dispersion coefficients for analyzing alternative residuals management strategies.

---

[25]The two plants are Gulf Oil and Atlantic Richfield, both in Philadelphia, and the receptor locations are in political jurisdictions 14, 15, 16, 17, 18, 19, and 20, all in Philadelphia. Thus, a total of 14 air dispersion coefficients are involved in this computation. These plants and receptor locations were selected for more precise treatment because: (1) both plants are large contributors to ambient ground level concentrations of sulfur dioxide, and (2) these plants are located close to an area in Philadelphia (especially receptor locations 14, 16, and 17) where ambient concentrations for all runs of the model are near, or substantially above, the federal primary standard for sulfur dioxide (80 μgms/m^3).

The technique that we used to select a single stack to represent a group of stacks involved determining stack characteristics that produced the same result, however defined, as the group of stacks taken together. We selected as the basis for our definition of "same" that the single stack should produce the same maximum ground level concentration as the group of individual stacks using discharge rates from EPA's inventory.[26] The discharge rate employed in the analysis for the single stack was the aggregate discharge of all the individual stacks. As noted above, this method does not, in general, ensure the same ground level concentrations throughout the area as would the group of stacks, except at the location of the maximum ground level concentration associated with the group of stacks.

Stack characteristics of the single stack (physical stack height, plume rise, and effective stack height) were determined for discharges of both sulfur dioxide and particulates. In general, the two "artificial" stacks computed on this basis differed substantially because the characteristics of the single stack depends on the relative discharge rates of all the individual stacks. In developing the two air dispersion matrices, with the exception of the two plants mentioned above, one set of single stacks was employed for the sulfur dioxide analysis, and a different set was employed for the suspended particulates analysis.[27]

---

[26] For groups of stacks with varying stack characteristics (stack height and plume rise), the stack characteristics of the single stack is a function of the distribution of discharges from the individual stacks.

[27] More information on the stack aggregation procedure is contained in an unpublished RFF memorandum [56].

V. Management Strategies Analyzed Using the Regional Model

We are currently in the process of making production runs using the Lower Delaware Valley model with various combinations of water quality and air quality "standards," and landfill quality requirements. For both air and water quality, we use two sets of standards: a set that is relatively easy to meet and a stricter set which, in the cases of sulfur dioxide and suspended particulates, is roughly equivalent to the federal primary standards. These standards are shown in Table 11. As will be demonstrated later on, it is technically possible to meet the water quality standards--with and without aerators. However, meeting the air quality standards appears to be a different matter. Using our model, the best air quality technically possible in some political jurisdictions barely meet the federal primary air quality standards for sulfur dioxide and suspended particulates.

For landfill operations, we use three quality levels: low, medium, and high. We define low quality as an open dump, though with no burning allowed; medium quality as a good quality sanitary landfill; and high quality as a good quality sanitary landfill with shredding, impervious layer to protect the groundwater, wastewater treatment of leachate, and protection of aesthetic qualities using fences, trees, and so on.

The extra cost of electricity, home and commercial heat, sewage treatment and disposal, and municipal solid residuals handling and disposal are not constrained in this set of production runs. That will be the subject of additional environmental policy analyses of this region.

More than 20 production runs have been made to explore the effects of the following policy variables: different requirements on the quality

Table 11

Air, Water, and Land Quality Constraints ("standards") for Production Runs of the Lower Delaware Valley Model

		Standard Sets	
		Easy to meet ("E")	More difficult to meet ("T")
Water Quality			
Algae	≤	3.0 mg/ℓ	2.0 mg/ℓ
Fish	≥	0.01 mg/ℓ	0.03 mg/ℓ
Dissolved oxygen	≥	3.0 mg/ℓ	5.0 mg/ℓ
Air Quality			
Sulfur dioxide	≤	120 μgms/m^3	80 μgms/m^3 *
Suspended particulates	≤	120 μgms/m^3	75 μgms/m^3 *

*For certain jurisdictions, it is technically impossible to meet the stricter standard, and we have had to make the indicated modifications:

Political jurisdiction	Sulfur dioxide	Suspended particulates
14		76 μgms/m^3
15		82 "
16	83 μgms/m^3	
22		76 "
23		79 "
24		79 "
27		81 "
50		76 "

Landfill Quality

L (low quality)	-- Open dump, but no burning allowed.
M (medium quality)	-- Good quality sanitary landfill.
H (high quality)	-- Good quality sanitary landfill with shredding, impervious layer to protect groundwater, wastewater treatment of leachate, aesthetic considerations such as fences, trees, etc.

of landfills; the in-stream aeration option for improving estuary water quality; the regional sewage treatment plant option; regional solid residuals management options, including incineration, railhaul out of the region, and recycling opportunities for used newsprint and corrugated containers; and different standards on the qualities of the estuary and of the air above the region. In this paper, we report on the production runs that are summarized in Table 12. The number of iterations of the regional model made for each run is also shown in this table.

In each run, we stopped after between 29 and 35 iterations, principally because of the expense of operating this model. At current computer rates, it costs $1,220 for a 30-iteration run, or roughly $41 per iteration.[28] Technically, these stopping points are not the optimum (even a local optimum) but our experience indicates that after this number of iterations the environmental constraints are met (if it is technically possible to meet them) and that continuing beyond this point improves the objective function value only slightly. This is not meant to imply, however, that residuals discharge patterns would not change--perhaps even substantially--if we continued beyond 30 iterations. A relevant question is how comparable are the results--costs and discharges--of the different production runs using output obtained from approximately the 30th iteration of each.[29]

---

[28]Currently we are operating on an IBM 370 Model 165 and are using 300 K bytes of internal core storage. Per iteration, it takes, on average, 2.80 minutes of CPU time, 5.94 minutes of I-O time, has 13,400 read-write instructions, and prints 1,236 lines of output (minimum).

[29]Although total discharges do not change much beyond the 30 to 35 iterations (because the ambient standards are being met), swapping among individual dischargers is still taking place. The one exception to the statement that total discharges do not change much beyond the 30th iteration is the tradeoff between in-stream aerators and BOD discharges.

Table 12

Characteristics of Selected Production Runs of the Lower Delaware Valley Model

Run number [c]	Standards [a]			Regional options [b]		Iterations of regional model
	Air	Water	Landfill	In-stream aeration	Sewage treatment	
1	T	T	H	+	0	34
2	T	E	H	+	0	35
3	E	T	H	+	0	29
4	E	E	H	+	0	30
5	T	T	M	+	0	35
6	T	E	M	+	0	34
7	E	T	M	+	0	30
8	E	E	M	+	0	30
9	T	0	H	+	0	35
10	0	E	H	+	0	30
11	0	T	H	+	0	16
12	E	0	H	+	0	30
13	E	T	H	0	0	30
14	E	E	H	0	0	30
15	E	T	H	0	+	29
16	E	T	H	+	+	30
17	∞	0	H	0	0	1
18	0	∞	H	0	0	1
19	∞	∞	H	0	0	1
23	0	∞	H	+	0	1
25	0	0	H	0	0	1
26	0	0	M	0	0	1
Base Case	0	0	L	0	0	1

For footnotes, see next page.

Footnotes to Table 12:

[a]Notation for standards. See Table 11.

Air and water quality:

T = most restrictive standards

E = easier to meet standards

O = no effective standard

∞ = extremely high marginal penalties on residuals discharges

Landfill quality:

H = high quality landfill

M = medium quality landfill

L = low quality landfill

[b]Notation for regional options:

\+ = regional option employed

O = regional option not considered

[c]Runs 20, 21, and 22 are similar to runs 17, 18, and 19 except that medium quality landfills are employed rather than requiring high quality landfills.

VI. Results Achieved to Date

The results achieved to date are organized and presented under the following subheadings: aggregate costs to the region; the distribution of costs; residuals discharges; solid residuals handling and disposal; and regional environmental quality.

Aggregate Costs to the Region

The objective function values for each of the six LP modules described in Table 6, and for the region as a whole, are presented in Table 13 for the initial set of production runs. These values represent the increased costs ($ per day) associated with improving the environmental quality of the region--atmosphere, estuary, and landfills--above the base case which is assumed to be the situation that existed in the Lower Delaware Valley during the period of roughly 1968 to 1970. The estimated costs reported here are based on the sum of the annual operating and maintenance costs for all residuals management activities in the region and the annualized capital cost of new construction using an 8 percent capital recovery factor. The base costs for the modeling studies were determined using the Lower Delaware Valley residuals management model with landfills of low quality throughout the region, and with no incentives to reduce either liquid or gaseous residuals discharges. For this analysis, effluent charges of zero were applied.[30] The objective function values

[30]At the optimum the marginal penalties are in fact the optimal set of effluent charges. Because the optimal sets of both residuals discharges and effluent charges are output of the analysis, either effluent standards or effluent charges could be used as the policy instrument for ensuring an efficient, or "optimal," management strategy.

Table 13

Objective Function Information for the Production Runs
--Lower Delaware Valley Model--

	Computer Run									
	1	2	3	4	5	6	7	8	9	10
Characteristics of runs:										
air standards	T	T	E	E	T	T	E	E	T	O
water standards	T	E	T	E	T	E	T	E	O	E
landfill standards	H	H	H	H	M	M	M	M	H	H
in-stream aeration	+	+	+	+	+	+	+	+	+	+
regional sewage treatment	O	O	O	O	O	O	O	O	O	O
Costs ($1,000 per day)[a]:										
Total regional:	1,201.0	1,179.0	425.0	362.0	1,164.0	1,139.0	371.0	312.0	1,119.0	109.0
LP Modules:										
MPSX 1	527.0	543.0	167.0	150.0	589.0	524.0	146.0	132.0	525.0	19.6
MPSX 2	463.0	455.0	84.6	81.0	395.0	465.0	88.0	87.7	462.0	0
MPSX 3	43.6	45.9	12.9	8.4	44.9	46.0	8.0	6.9	46.5	0
MPSX 4	53.4	31.8	53.3	38.4	47.7	28.6	50.9	29.5	0.3	35.8
MPSX 5	78.7	79.0	71.4	60.0	55.2	54.4	39.9	37.5	79.4	35.7
MPSX 6	35.5	24.9	36.2	23.8	32.8	20.0	37.8	18.9	5.5	17.9

For footnote, see next page.

continued ...

Table 13 (continued)

	Computer Run									
	11	12	13	14	15	16	17	18	19	23
Characteristics of runs:										
air standards	0	E	E	E	E	E	∞	0	∞	0
water standards	T	0	T	E	T	T	0	∞	∞	∞
landfill standards	H	H	H	H	H	H	H	H	H	H
in-stream aeration	+	+	0	0	0	+	0	0	0	+
regional sewage treatment	0	0	0	0	+	+	0	0	0	0
Costs ($1,000 per day)[a]:										
Total regional:	145.0	265.0	519.0	351.0	463.0	393.0	1,193.0	343.0	1,411.0	357.0
LP Modules:										
MPSX 1	24.4	113.0	184.0	136.0	139.0	150.0	586.0	86.0	584.0	86.0
MPSX 2	0	84.8	95.4	86.0	89.5	85.1	474.0	0	474.0	0
MPSX 3	0	7.3	9.0	7.8	8.0	8.8	47.2	0	50.0	0
MPSX 4	50.4	0	107.0	42.2	111.0	50.1	0.3	156.0	157.0	156.0
MPSX 5	36.3	56.3	64.4	59.7	62.0	62.0	79.5	39.3	81.6	39.3
MPSX 6	34.2	3.5	59.1	18.7	53.3	36.6	5.5	63.8	69.2	77.5

[a]These costs represent the increased costs associated with improving the environmental quality of the Lower Delaware Valley region above the base case described in the text. These costs are based on the sum of the annual operating and maintenance costs and the annualized capital costs of new construction using an 8 percent capital recovery factor.

reported in Table 13 are net of the base costs. In addition, they represent only the costs; the calculated penalties required by the optimization algorithm are not included. The reported objective function value for a situation with low landfill quality and with no effluent charges (marginal penalties) on either liquid or gaseous emissions would be zero.[31]

To examine the impact on the total costs to the region of meeting different sets of air quality standards, different sets of water quality standards, and different sets of air and water quality sta dards considered simultaneously, Figure 7 has been prepared using the cost data from Table 13. (The cost information contained in this 16-element matrix is associated with high landfill quality.) There are four entries in this table that are outputs of single iteration computer runs--runs 17, 18, 19, and 25. These four costs are not dependent on the nonlinear optimization algorithm.

As one would expect, total costs to the region increase as the air and water quality standards become more stringent. It is interesting to note in this figure that the total cost to the region of meeting "T" water quality standards ($145,000 per day) is substantially less than the costs of meeting "T" air quality standards ($1,119,000 per day). This is due in part to the fact that we have considered most of the gaseous residual discharges in the region, but have considered only a portion of the liquid residuals dischargers; specifically, only those that discharge their waterborne residuals to the Delaware Estuary. Nevertheless, it

---

[31]For high (H) and medium (M) quality landfills, but with no air or water quality constraints imposed--computer runs 25 and 26, respectively--the total regional costs are $33,769 and $16,020 per day, respectively. These costs are all attributable to Module 5 (MPSX 5).

Air Quality Standards

Water Quality Standards	0	E	T	∞
0	$33,800[a,b] (#25)	$265,000 (#12)	$1,119,000 (#9)	$1,193,000[a] (#17)
E	$109,000 (#10)	$362,000 (#4)	$1,179,000 (#2)	
T	$145,000 (#11)	$425,000 (#3)	$1,201,000 (#1)	
∞	$343,000[a] (#18)			$1,411,000[a] (#19)

Notes: The numbers in parentheses indicate the computer run number (see Table 12).

The data reported in this figure are based on computer runs where high (H) quality landfills are required.

[a]Single iteration runs. Aerators are not employed in these runs.

[b]The cost reported here is for a high (H) quality landfill run. It should not be confused with the base case where regional costs are, in fact, zero. See Table 12 (p. 54) and footnote 31 (p. 59).

Figure 7. Total increased costs to the Lower Delaware Valley Region of meeting different sets of air and water quality standards--high quality landfills. Costs are in dollars per day.

appears that cleaning up the air will be a more pervasive and costly problem for the region than cleaning up the water.

With regard to the regional options considered in the model--in-stream aeration and regional sewage treatment plants--it appears to be advantageous for the region to employ both. To examine the feasibility of these options, we compare in Figure 8 computer runs 3, 13, 15, and 16, all having restrictive water quality standards (T), less restrictive air quality standards (E), and high quality landfills (H). With neither the in-stream aeration nor regional sewage treatment plant options considered (run 13), the total regional costs are $519,000 per day; with regional treatment but no in-stream aeration, the total regional costs amount to $463,000 per day (run 15); with in-stream aeration but no regional treatment, the costs drop to $425,000 per day (run 3); and with both regional treatment and in-stream aeration, the total regional costs drop still further to $393,000 per day (run 16). This result is consistent with the findings of other investigators [51,54].[32]

Also shown in Figure 8 are the BOD discharges from the municipal sewage treatment plants and the total aerator horsepower throughout the estuary for the four combinations of in-stream aeration and regional sewage treatment plants presented above. We note that the in-stream aeration option allows a substantial increase in the BOD discharges; also, that without in-stream aeration the total discharges of BOD from the

[32]The regional treatment plant scheme does, however, pose an interesting distributional problem. In comparing runs 13 and 15 (Table 13) where no in-stream aeration is provided, note that the total cost of municipal sewage treatment (MPSX 4) actually increases when regional treatment plants are introduced. The savings in costs accrue to the industrial dischargers (MPSX 6).

Regional sewage treatment	In-stream aeration 0	In-stream aeration +
0	$519 (#13)	$425 (#3)
+	$463 (#15)	$393 (#16)

Total regional costs[a] ($1,000 per day)

Regional sewage treatment	In-stream aeration 0	In-stream aeration +
0	77.4 (#13)	123.8 (#3)
+	66.8 (#15)	131.7 (#16)

BOD discharges from municipal sewage treatment plants[b] (1,000 lbs/day)

Regional sewage treatment	In-stream aeration 0	In-stream aeration +
0	0 (#13)	14.3 (#3)
+	0 (#15)	11.1 (#16)

Aerator horsepower[b] (1,000 horsepower)

Notes: Numbers in parentheses indicate the computer run.

The data reported in this figure are based on computer runs where the following conditions are imposed: E level air quality standards, T level water quality standards, and H level landfill quality restrictions.

[a]Taken from Table 13.

[b]Taken from Table 16.

Figure 8. Comparison of costs and discharges for regional options--in-stream aeration and regional sewage treatment.

municipal treatment plants is smaller for the regional plants than for the individual plants due to the more extensive distribution of discharges, and hence more efficient use of the estuary's assimilative capacity, from the latter.

The alert reader will note in Table 13 that under the less restrictive (E) water quality standards there is an apparent contradiction. With the aerator option allowed (run 4), the aggregate cost to the region is $362,000 per day, while when the option is removed (run 14), the cost drops to $351,000 per day. This appears at first glance to be evidence of multiple optima, since clearly in run 4 the model could have chosen <u>not</u> to use the aerators. On closer examination, however, we find that this is simply an artifact of our arbitrary strategy involving stopping at about the 30th iteration. Had we continued the interative procedure (for run 4) far enough, the aerator activity levels would have eventually returned to zero. In the beginning of the process, aerators come in to aid in meeting the dissolved oxygen standards. However, after these standards have been met (around the 10th iteration), the aerators start trading off with other residuals discharges, most notably BOD. From about the 20th iteration on, both the total aerator activity level and the total BOD discharges are progressively being reduced. The step sizes of the aerators and residuals discharges around the 30th iteration are relatively short, and hence the process still has a long way to go before the aerators would actually reach the zero activity level. Note also the relatively flat response surface. The difference between the total regional cost of runs 4 and 14 is only about $11,000 per day, or about 3 percent of the total regional costs of improving the quality of the environment.

The finding that total regional costs are greater with aerators than without for the less restrictive water quality standards (E) was somewhat of a surprise. One possible explanation for this outcome is that at the higher levels of residuals discharges associated with E level ambient standards (lower levels of removal), the marginal costs of treatment are less than the marginal costs of employing aerators. In comparing wastewater discharges for the relevant runs, the average BOD removal efficiency for all estuary dischargers required to meet the E level water quality standard is about 58 percent with no aerators (run 14) and 52 percent with aerators (run 4); for the municipal dischargers, the removal efficiencies are 65 and 58 percent, respectively.[33] For the T level water quality standards, the comparable BOD removal efficiencies are summarized in the following table:[34]

Category of Discharger	With aerators (run 3) (%)	Without aerators (run 13) (%)
All estuary dischargers	71	84
Municipal dischargers	73	83

We see from this comparison that the BOD removal efficiencies for the E level water quality standards are substantially smaller than those associated with the T level standards. These rather large differences in the removal efficiencies imply substantially smaller marginal costs of wastewater treatment in the case of meeting the E level water quality standards.

[33] These removal efficiencies are based on the wastewater discharge data presented in Table 16 using as the basis of the computations the BOD discharges of run 25.

[34] The removal efficiencies in this table are based on the wastewater discharge data presented in Table 16 using as the basis of the computations the BOD discharges of run 25.

In comparing regional costs of meeting the E level water quality standards with and without aerators (Table 13, computer runs 4 and 14), we note that without aerators the cost of municipal sewage treatment (MPSX 4) increase $3,800 per day and the costs of industrial wastewater treatment (MPSX 6) increase $200 per day for a total of $4,000 per day for these two categories of dischargers. However, for this management strategy, the aerator costs drop $5,300 per day,[35] and thus, based on a partial analysis using only these two discharge categories, this strategy represents a net savings to the region of $1,300 per day.

These cost differences suggest that even though the marginal costs of wastewater treatment appear to be less than the marginal costs of in-stream aeration at the E level water quality standards, the differences are very small indeed. From an overall regional cost point of view, it appears that either option could be employed without much of an impact on regional efficiency.

To examine the impact on the costs to the region of different landfill qualities and different sets of air and water quality standards, Figure 9 has been prepared using the data from runs 1 through 8. In this figure, total regional costs and the costs of paper production and solid residuals handling and disposal (Module 5) are compared for medium quality landfills (M) and high quality landfills (H), and for various combinations of E and T level air and water quality standards. From this information we can see that the total incremental costs to the region of going from medium quality landfills (M) to high quality landfills (H) amounts to

---

[35]The cost savings associated with not using aerators ($5,300) is based on cost information presented in Table 14 (run 4).

Total regional costs:

High quality landfills

Water quality standards	Air quality standards: E	Air quality standards: T
E	$362 (#4)	$1,179 (#2)
T	$425 (#3)	$1,201 (#1)

Medium quality landfills

Water quality standards	Air quality standards: E	Air quality standards: T
E	$312 (#8)	$1,139 (#6)
T	$371 (#7)	$1,164 (#5)

Differences in costs

Water quality standards	Air quality standards: E	Air quality standards: T
E	$50	$40
T	$54	$37

Module 5 costs: Paper plants and municipal solid residuals handling and disposal activities

High quality landfills

Water quality standards	Air quality standards: E	Air quality standards: T
E	$60 (#4)	$79 (#2)
T	$71.4 (#3)	$78.7 (#1)

Medium quality landfills

Water quality standards	Air quality standards: E	Air quality standards: T
E	$37.5 (#8)	$54.4 (#6)
T	$39.9 (#7)	$55.2 (#5)

Differences in costs

Water quality standards	Air quality standards: E	Air quality standards: T
E	$22.5	$24.6
T	$31.5	$23.5

Note: Numbers in parentheses indicate computer runs (see Table 12).

Figure 9. Comparison of costs for high and medium quality landfills ($1,000 per day)

somewhere between \$40,000 and \$50,000 per day. However, only about \$23,000 per day is attributable to the paper plant-solid residuals management module. Thus, landfill quality requirements impose substantial incremental costs on other, apparently unrelated, activities in the region. The second thing to note from Figure 9 is that the water quality standards seem to have very little impact on the costs of paper production and municipal solid residuals management (Module 5), whereas the air quality standards appear to have a substantial impact on costs.

The Distribution of Costs

In Table 14, we present information on the increases in certain consumer costs implied by improved regional environmental quality for 14 selected model runs. Specifically, this includes, for the 57 political jurisdictions of the region, the minimum, mean, and maximum percent increases in costs of household electricity, home and commercial space heating, and solid waste disposal; and the minimum, mean, and maximum increased costs of sewage disposal (\$ per household per year) and in-stream aeration (\$ per political jurisdiction per day). The jurisdictions experiencing the minimum and maximum increases are indicated in parentheses. In some cases, more than one political jurisdiction is at the maximum value, although only one jurisdiction is indicated; similarly for the minimum value. The runs displayed are: the first eight with the permutations of T and E level water and air quality standards, and high and medium landfill quality requirements; 13, 15, and 16, which involve E level air quality standards, T level water quality standards, high landfill quality, and various combinations of the in-stream aeration and regional treatment plant options; 19, which involves high effluent charges

Table 14

Distribution of Increased Consumer Costs for the Lower Delaware Valley Region

		Computer Run						
		1	2	3	4	5	6	7
Characteristics of runs:								
air standards		T	T	E	E	T	T	E
water standards		T	E	T	E	T	E	T
landfill standards		H	H	H	H	M	M	M
in-stream aeration		+	+	+	+	+	+	+
regional sewage treatment		0	0	0	0	0	0	0
Electricity:[a] (percent)	min	1.0 (49)	1.4 (49)	0.2 (49)	0.1 (49)	1.4 (49)	1.1 (49)	0.1 (49)
	mean	11.3	11.2	3.3	2.8	10.4	11.3	3.0
	max	15.8 (4)	15.4 (4)	4.5 (42)	3.7 (42)	15.4 (4)	16.0 (4)	4.1 (42)
Commercial heat:[b] (percent)	min	67.8 (39)	39.3 (46)	0 (46)	0 (57)	0 (46)	109.2 (57)	0 (46)
	mean	115.6	114.1	24.7	22.6	102.5	117.8	23.5
	max	117.8 (23)	117.8 (23)	105.4 (16)	117.8 (18)	117.8 (23)	117.8 (23)	117.8 (17)
Home heat: (percent)	min	9.9 (32)	9.9 (32)	0 (57)	0 (57)	0.4 (3)	4.9 (57)	0 (57)
	mean	26.8	25.8	4.2	4.1	21.8	26.4	4.7
	max	55.4 (23)	55.4 (23)	30.0 (23)	43.8 (19)	55.4 (23)	55.4 (23)	46.6 (19)
Sewage disposal:[c] ($/household/ year)	min	0.5 (1)	0 (49)	0.5 (1)	0 (49)	0.4 (1)	0 (44)	0.5 (1)
	mean	13.4	7.9	13.3	9.5	11.6	7.1	12.5
	max	48.2 (44)	36.8 (50)	49.2 (50)	37.2 (50)	51.2 (50)	38.7 (50)	46.7 (2)
Solid waste disposal: (percent)	min	16.6 (24)	16.6 (24)	7.6 (15)	7.8 (15)	6.6 (24)	6.6 (24)	6.6 (24)
	mean	19.2	19.6	18.0	18.6	7.6	7.6	7.7
	max	21.1 (26)	23.6 (32)	23.1 (26)	24.9 (32)	7.8 (57)	7.8 (57)	8.2 (26)
In-stream aeration:[d] ($/jurisdiction/day)		141.8	89.4	156.6	92.7	110.1	65.9	131.1

Note: Numbers in parentheses indicate the jurisdiction with the minimum and maximum increased cost.

For footnotes, see next page.

continued ...

Table 14 (continued)

		Computer Run						
		8	13	15	16	19	25	26
Characteristics of runs:								
air standards		E	E	E	E	∞		
water standards		E	T	T	T	∞		
landfill standards		M	H	H	H	H		
in-stream aeration		+	0	0	+	0		
regional sewage treatment		0	0	+	+	0		
Electricity:[a] (percent)	min	0.1 (49)	0.2 (49)	0 (49)	0.2 (49)	1.1 (49)		
	mean	2.3	3.1	2.4	3.0	11.5	0	0
	max	3.1 (42)	4.1 (42)	3.2 (42)	3.9 (42)	16.3 (4)		
Commercial heat:[b] (percent)	min	0 (49)	0 (46)	0 (57)	0 (57)	117.8		
	mean	25.1	25.8	24.9	23.0	117.8	0	0
	max	117.8 (18)	117.8 (18)	117.8 (18)	117.8 (18)	117.8		
Home heat: (percent)	min	0 (57)	0 (57)	0 (57)	0 (57)	9.9 (32)		
	mean	4.2	4.4	4.8	4.4	27.4	0	0
	max	42.5 (19)	44.6 (19)	42.0 (19)	36.9 (19)	55.4 (23)		
Sewage disposal:[c] ($/household/ year)	min	0 (49)	1.5 (1)	0 (57)	0 (20)	1.5 (1)		
	mean	7.3	25.9	26.3	11.2	36.3	0	0
	max	40.3 (50)	139.5 (2)	124.8 (2)	47.0 (2)	139.5 (2)		
Solid waste disposal: (percent)	min	6.6 (24)	7.8 (15)	7.7 (15)	7.7 (15)	16.6 (24)	9.2 (15)	7.6 (32)
	mean	7.7	17.9	17.9	18.0	19.1	16.2	7.8
	max	9.0 (26)	24.1 (32)	23.2 (26)	21.1 (26)	19.6 (57)	22.5 (25)	7.9 (57)
In-stream aeration:[d] ($/jurisdiction/day)		60.4	0	0	121.9	0	0	0

Note: Numbers in parentheses indicate the jurisdiction with the minimum and maximum increased cost.

[a]Does not include jurisdiction 39 which is always 0.

[b]Does not include jurisdictions 19, 28, 29, 31, 32, 33, 37, and 44 which are always 0.

[c]Does not include jurisdictions 5, 6, 7, 36, 43, 45, 46, 53, 54, and 56 which are always 0.

[d]Total average daily costs divided by 57.

on discharges of both liquid and gaseous residuals and thus provides a set of upper bounds on increased consumer costs; and 25 and 26, which involve no effluent charges on liquid and gaseous residuals, for the high and medium quality landfills, respectively.

The results shown do not, of course, capture the full complexity of the geographical distribution of these costs throughout the region. For the reader interested in more detail on the distributional implications, we provide data in Appendix C (Table C-1) on the distributional consequences of a variety of runs:

(1) a comparison of increased home heating costs for two situations:
 (i) restrictive air quality standards (run 1), and
 (ii) less restrictive air quality standards (run 3);

(2) a comparison of increased household electricity costs for the same two situations as for home heating;

(3) a comparison of increased municipal solid wastes handling and disposal costs for two situations:
 (i) high quality landfills with less restrictive air quality standards (run 3), and
 (ii) medium quality landfills with less restrictive air quality standards (run 7);

(4) a comparison of the increased costs of municipal sewage disposal for two situations:
 (i) restrictive water quality standards (run 1), and
 (ii) less restrictive water quality standards (run 2).

These results will be referred to in the text as appropriate.

As shown in Table 14, the maximum increase in electricity bills in an individual jurisdiction when tight air quality standards are imposed is about 16 percent. The most interesting thing to note about this is that the maximum increase did not occur in Philadelphia, where one might have expected it to, but in New Castle County, Delaware, jurisdictions 1, 2, 3, and 4 (Table C-1).[36] The highest regional average increase in electricity bills amounts to 11.3 percent (run 1). This would appear to demonstrate that improved air and water quality in the Lower Delaware Valley region will have a relatively small impact on costs of electricity generated within the region, though with electricity bills already soaring due to increases in fuel prices, the increase due to improve air quality may cause a straw-that-broke-the-camel's-back effect.

Information on increases in costs of commercial heat is included in Table 14 only. The maximum value obtainable is 117.8 percent. This represents 100 percent conversion to natural gas. This maximum value is obtained for at least one jurisdiction (in Philadelphia) for almost all production runs with either E or T level air quality standards. The average increase for the region is relatively high (over 100 percent) for all runs with T level air quality standards, while for E level air quality standards, the average increase is on the order of only 25 percent. Similarly, the minimum observed increases in commercial heating costs vary significantly with the air quality standard imposed, being over 100 percent under some conditions and zero under others.

---

[36]It will be noted by those who examine Table C-1 that large groups of political jurisdictions move together in their electricity cost increases. This is because there are only 4 utilities serving the region (with a few minor exceptions of locally owned generating stations and of jurisdictions on the regional boundary which are partly served by other companies with no generating plants in the region) [31(Appendix D),57]. The increased cost implied for each utility as a whole has been applied equally to all households in its service area.

For home heat, the maximum possible increase in costs (associated with 100 percent conversion to natural gas) varies among the jurisdictions depending on the distribution of fuels in use in 1970 for home-heating purposes. The largest such number is 55.4 percent in political jurisdiction 23, and this is the observed regional maximum for the tight air quality runs shown in Table 14. When the less restrictive air quality standards are imposed, the maximum observed increased costs do not reach this highest possible level, but are still large, running on the order of 35 to 45 percent. Proportionally, the observed average and minimum cost increases are even more sensitive to the choice of air quality standards, varying between 20-25 percent and 4-5 percent for the former, and 10 and zero percent for the latter, when the standards are relaxed from T to E level.

The maximum observed increase in sewage disposal bills for the region is about $140 per household per year in jurisdiction 2--Wilmington, Delaware--(run 13). This cost increase occurs when neither in-stream aeration nor regional sewage treatment is provided as an option. When regional treatment plants are provided (run 15), the maximum increase in sewage disposal bills is still associated with jurisdiction 2, but the amount drops to about $125, indicating that less treatment is required at the Wilmington sewage treatment plant. With in-stream aeration provided but no regional treatment plants (run 3), the maximum increase is about $50 in jurisdiction 50--Camden, New Jersey. (For this case, political jurisdiction 2 drops to the number two position at about $47 per household per year.) For the case with both in-stream aeration and regional sewage treatment provided (run 16), jurisdiction 2 returns to the position of maximum increase in sewage disposal bills with a cost increase of \$47 per household per year. Thus, at least for political jurisdiction 2, the district associated with the maximum increase in sewage disposal costs for many of the production runs, regional options--in-stream aeration and sewage treatment--have little impact on shifting its relative cost position among jurisdictions, though they do affect its absolute dollar burden.[37] This is probably not true, however, of some of the other political jurisdictions. One expects that, although regional water quality management options are generally less expensive for the region as a whole, without constraining the distribution of costs the burden will shift among jurisdictions. This will be examined in more detail in the section on distributional implications.

---

[37] Of course, when in-stream aeration is provided, the total dollar burden to a community depends both on the costs of sewage treatment and on the costs of in-stream aeration, and in particular on how the costs of in-stream aeration are allocated between the industrial and municipal sectors, and then for the municipal sector, among the communities within the region. One possibility would be to distribute the costs of in-stream aeration in proportion to the saving in treatment costs of each discharger made possible by the aeration option. Another possibility would be to distribute the aeration costs in proportion to the reduction in BOD discharges. Based on the latter consideration, the following costs would obtain for Wilmington (jurisdiction 2) and Camden (jurisdiction 50) for runs 3 and 16 of the regional model:

Costs of In-stream Aeration
(\$ per household per year)

Name of City	Political jurisdiction	Computer Run	
		Run 3[a]	Run 16[b]
Wilmington	2	\$1.60	\$1.20
Camden	50	0.90	2.70

[a] Based on BOD reductions due only to in-stream aeration.

[b] Based on BOD reductions due to both in-stream aeration and the regional treatment plants.

In Table C-1, a comparison is made of the distribution of increased municipal sewage disposal costs for E and T level water quality standards (runs 1 and 2). Here, the variation among political jurisdictions is substantial, and the jurisdictions experiencing the minimum and maximum increases under the two environmental quality conditions change. We can also see, from the data shown in Table 14, that the observed maximum, average, and minimum of these increased costs are not particularly sensitive to the level of water quality required, though the average can vary by a factor of nearly 2 in going from the T level to the E level standards.

As can be seen from Table C-1, for a given run the variation in increased costs of municipal solid waste disposal over political jurisdictions is small. This is because for each run all jurisdictions were required to have the same quality landfills. (We did not investigate the impact of variations in landfill quality among jurisdictions.) The variation in increased costs among jurisdictions for the high quality landfills (run 3) is greater than for medium costs landfills (run 7) because of the differences in the costs among jurisdictions of alternatives to disposal in landfills. For high quality landfills, the average increased costs are in the order of 18 percent (the maximum about 25 percent); for medium quality landfills, the average increase drops to less than 8 percent.

Residuals Discharges

In Table 15, we present, for some of the production runs, the total gaseous discharges of sulfur dioxide and particulates for selected groups of activities in the region: petroleum refineries, steel mills, electric power plants, home and commercial heating, the group of industries in the "over 25 $\mu gms/m^3$" category (Table 6), paper plants, and municipal

Table 15

Gaseous Discharges for Selected Groups of Dischargers
(tons per day)

Discharger Group	Computer Run								
	1	2	3	4	5	6	7	8	9
Air Quality Standards	T	T	E	E	T	T	E	E	T
Sulfur dioxide:									
Petroleum refineries	185	185	322	326	199	185	337	325	184
Steel mills	10	5	92	88	8	5	90	87	6
Electric power plants	6	6	485	574	21	10	610	637	18
Home and commercial heat	19	25	494	527	83	6	510	507	6
"Over 25 μgms/m^3" dischargers	39	38	99	114	0	37	113	121	36
Paper plants	56	56	72	104	56	56	101	106	56
Total	318	317	1,572	1,743	410	301	1,771	1,792	307
Particulates:									
Petroleum refineries	2.7	2.7	41.2	43.7	5.4	2.7	67.1	62.6	2.7
Steel mills	12.9	12.8	80.0	82.9	14.1	12.8	83.8	84.6	14.6
Electric power plants	3.1	4.0	100.3	100.0	8.3	6.1	107.7	105.7	4.7
Home and commercial heat	9.4	10.9	53.5	52.9	18.5	9.1	52.1	52.5	9.2
"Over 25 μgms/m^3" dischargers	2.7	2.5	37.5	37.1	2.6	2.4	45.8	43.7	2.5
Municipal incinerators	0	0	31.7	34.3	0	0	0	0	0
Total	37.6	39.6	354.2	360.5	55.7	39.9	364.9	358.2	40.4

continued ...

Table 15 (continued)

Discharger Group	Computer Run							Maximum Reduction Efficiency[b]
	10	11	12	17	18	19	25[a]	
Air Quality Standards	0	0	E	∞	0	∞	0	
Sulfur dioxide:								
Petroleum refineries	457	457	334	183	457	183	457	60
Steel mills	93	93	84	4	53	4	93	96
Electric power plants	2,100	1,176	647	3	1,598	3	1,569	99.8
Home and commercial heat	779	779	525	0	779	0	779	100
"Over 25 μgms/m^3" dischargers	275	275	119	35	275	35	275	87
Paper plants	191	190	111	56	189	56	192	71
Total	3,920	3,596	1,828	282	3,376	283	3,390	92
Particulates:								
Petroleum refineries	84.8	84.8	45.0	2.7	84.8	2.7	84.8	97
Steel mills	84.2	84.2	84.4	12.2	18.9	12.9	84.2	86
Electric power plants	145.2	144.0	102.8	2.2	142.1	2.2	151.1	98.5
Home and commercial heat	58.1	58.1	52.5	8.0	58.1	8.0	58.1	86
"Over 25 μgms/m^3" dischargers	73.6	73.6	36.2	2.6	73.6	2.6	73.6	96
Municipal incinerators	72.8	72.8	36.1	0	72.8	0	70.5	100
Total	528.3	527.5	366.0	34.2	460.7	35.0	531.3	94

[a] Run 25 represents high quality landfill (H) and zero effluent charges on both gaseous and liquid residuals.

[b] Maximum reduction efficiency = $\frac{\text{Run 25} - \text{Run 17}}{\text{Run 25}}$.

incinerators. The differences between total gaseous discharges for the E and T level ambient air quality standards are quite substantial for every category of discharger and for both sulfur dioxide and particulates discharges. This difference is most dramatic for the power plants and the space heating activities (for sulfur dioxide in the latter case). The refineries and paper plants do proportionally less cutting back in SO_2 discharges. Nothing in the table, of course, matches the sensitivity of the incinerators to the standards placed on particulate discharges--the strict standards drive the incinerators right out of the optimal solution.[38]

Also shown in Table 15 are the maximum removal efficiencies for selected groups of dischargers. As one can see, these efficiencies are relatively high. The averages for all dischargers in the model are 92 percent for sulfur dioxide and 94 percent for particulates. The very high reduction levels for power plants, and home and commercial heating, reflect our assumption, already discussed, that a complete switch to natural gas is possible for these operations. The rather low figure for maximum SO_2 reduction at the refineries is the result of two factors. First, our best estimate of the actual maximum capability of refineries to reduce SO_2 discharges, without converting completely to natural gas, but being able to replace almost all purchased fuel with desulfurized refinery gases, is about 75 percent [13]. For the Delaware regional model, in addition, some initial tuning of our refinery models was undertaken to bring our calculated discharges into line with those shown in EPA's inventory

---

[38]Neither are the incinerators used when landfill restrictions are eased, even when the air quality standards are also set at the E level (runs 7 and 8).

of gaseous emissions. This reduced the base discharges but not the attainable minimum for the modeled plants; hence, the 60 percent maximum reduction efficiency.

In the aggregate, the approximate discharge reductions required to achieve the required environmental quality levels throughout the region may be summarized as follows:[39]

Air Quality Level	Discharge Reduction Efficiency (%)	
	SO_2	Particulates
T	91	93
E	48	33

Table 16 shows, for some production runs, the BOD and suspended solids (SS) discharged by selected groups of activities in the region: petroleum refineries, steel mills, municipal sewage treatment plants discharging to the estuary, paper plants, and industrial dischargers to the estuary. In addition, we present heat discharges (Btu) from the petroleum refineries, steel mills, and electric power plants in the region, and the total horsepower of the aerators along the 22 reaches of the estuary.

As with the gaseous discharges, the variation in liquid discharges and aeration horsepower among computer runs is substantial. The maximum attainable removal efficiencies for BOD, suspended solids, and heat are all relatively high, being, in aggregate, 91 percent for BOD, 93 percent for suspended solids, and 100 percent for heat. For the maximum reduction case (runs 18 and 19), the petroleum refineries are particularly efficient

[39]These efficiencies are based on the high (H) quality landfill runs (runs 1, 2, 3, 4, and 25) using as the basis of the computations the gaseous discharges of run 25.

Table 16

Wastewater Discharges for Selected Groups of Dischargers

	Computer Run						
	1	2	3	4	5	6	7
Water Quality Standards	T	E	T	E	T	E	T
BOD (1,000 lbs/day):							
Petroleum refineries	14.0	11.2	28.1	46.5	11.4	7.9	29.9
Steel mills	2.4	2.6	3.2	3.1	3.0	2.5	2.8
Municipal sewage treatment	125.1	224.1	123.8	190.6	149.5	242.6	125.7
Paper plants	35.9	41.6	25.7	49.4	26.5	29.1	23.7
Delaware Estuary industries	83.5	143.6	76.8	138.2	86.5	152.9	75.7
Total	260.9	423.1	257.6	427.8	276.9	434.9	257.8
SUSPENDED SOLIDS (1,000 lbs/day):							
Steel mills	241.6	249.0	164.2	183.8	259.8	247.6	171.1
Municipal sewage treatment	220.4	454.8	219.0	373.1	277.2	503.9	220.3
Paper plants	125.9	138.7	116.4	160.6	112.9	126.7	112.7
Delaware Estuary industries	52.0	66.5	37.5	65.2	45.3	74.2	36.7
Total	653.1	909.1	538.3	782.8	695.1	952.4	540.9
HEAT (10^9 Btu/day):							
Petroleum refineries	122.9	101.6	155.7	158.9	112.7	110.5	160.5
Steel mills	29.7	24.3	5.6	10.2	29.9	24.4	9.8
Electric power plants	209.9	347.1	197.9	304.2	157.6	374.9	124.5
Total	362.5	472.9	359.2	473.3	300.1	509.8	294.9
OXYGEN							
Aerators (1,000 horsepower)	12.9	8.2	14.3	8.5	10.0	6.0	12.0

continued ...

Table 16 (continued)

	Computer Run						
	8	9	10	11	12	13	14
Water Quality Standards	E	O	E	T	O	T	E
BOD (1,000 lbs/day):							
Petroleum refineries	43.7	2.5	35.9	19.4	66.4	13.9	46.9
Steel mills	3.6	10.0	3.0	3.1	3.4	1.2	3.2
Municipal sewage treatment	233.5	452.8	206.6	126.5	452.8	77.4	158.8
Paper plants	35.3	26.9	50.0	26.0	122.3	18.7	49.0
Delaware Estuary industries	141.6	231.0	142.6	76.9	231.0	31.0	111.4
Total	457.6	723.3	438.1	251.9	875.9	142.1	369.4
SUSPENDED SOLIDS (1,000 lbs/day):							
Steel mills	165.5	249.2	184.9	101.9	177.9	191.5	173.7
Municipal sewage treatment	490.0	1,006.0	414.0	224.7	1,006.0	91.3	297.0
Paper plants	140.3	125.9	162.0	114.3	254.8	102.4	158.9
Delaware Estuary industries	72.3	117.3	67.1	38.0	117.3	4.7	56.9
Total	868.1	1,498.5	828.0	478.9	1,556.0	390.8	687.5
HEAT (10^9 Btu/day):							
Petroleum refineries	229.6	46.2	140.0	128.6	272.5	66.9	220.8
Steel mills	26.1	86.8	10.4	7.0	29.5	23.6	14.2
Electric power plants	409.8	1,283.8	311.9	164.4	825.1	20.6	213.6
Total	665.6	1,416.8	462.3	300.0	1,127.0	111.1	448.6
OXYGEN							
Aerators (1,000 horsepower)	5.5	0	7.5	13.0	0	0	0

continued ...

Table 16 (continued)

	Computer Run						Maximum Reduction Efficiency[b]
	15	16	17	18	19	25[a]	
Water Quality Standards	T	T	0	∞	∞	0	
BOD (1,000 lbs/day):							
Petroleum refineries	19.1	29.6	0.0	0.0	0.0	70.3	100
Steel mills	2.2	3.2	2.5	1.2	1.2	3.2	62
Municipal sewage treatment	66.8	131.7	452.8	39.4	39.4	452.8	91
Paper plants	25.1	27.1	20.9	14.7	14.7	130.0	89
Delaware Estuary industries	37.5	75.3	231.0	27.2	27.2	231.0	88
Total	150.7	266.8	707.2	82.5	82.5	887.3	91
SUSPENDED SOLIDS (1,000 lbs/day):							
Steel mills	171.1	180.6	37.4	6.1	6.1	22.8[c]	73
Municipal sewage treatment	60.1	244.1	1,006.0	2.0	2.0	1,006.0	99.8
Paper plants	110.8	110.5	117.4	85.5	85.5	265.7	68
Delaware Estuary industries	11.8	38.9	117.3	0.3	0.3	117.3	99.7
Total	355.4	575.4	1,278.0	93.9	93.9	1,411.8	93
HEAT (10^9 Btu/day):							
Petroleum refineries	149.2	151.9	0.0	0.0	0.0	272.5	100
Steel mills	27.4	6.5	8.5	0.0	0.0	28.1	100
Electric power plants	128.9	160.8	568.6	0.0	0.0	619.9	100
Total	305.6	319.2	577.2	0.0	0.0	920.5	100
OXYGEN							
Aerators (1,000 horsepower)	0	11.1	0	0	0	0	

[a] Run 25 represents high quality landfill (H) and zero effluent charges on gaseous and liquid residuals.

[b] Maximum reduction efficiency = $\frac{\text{Run 25 - Run 18}}{\text{Run 25}}$.

[c] Only 2 steel mills discharge to the estuary and hence have effluent charges applied to their waterborne discharges. The higher-than-base-case suspended solids loads in many runs may be artifacts of our limited aquatic modeling.

in removing BOD (100 percent), and the municipal sewage treatment plants and estuary industrial dischargers are extremely efficient in removing suspended solids (99.7 and 99.8 percent, respectively). The petroleum refineries, steel mills, and electric power plants can all eliminate completely their heat discharges to the water.

Actual discharge reduction efficiencies required in the aggregate to meet the two levels of ambient water quality standards are summarized below for the situation in which T level air quality and high landfill quality are required, and in which the aeration option is available (runs 3 and 4).

Water Quality Level	Discharge Reduction Efficiency (%)		
	BOD	SS	Heat
T	71	62	61
E	52	44	49

It is also interesting to observe how the actual aggregate discharge reduction efficiencies for BOD, suspended solids, and heat vary with the availability of the regional options--in-stream aeration and regional sewage treatment plants--for the T level water quality standards (runs 3, 13, 16, and 25).

Regional Management Options	Discharge Reduction Efficiency (%)		
	BOD	SS	Heat
Without aerators or regional plants (run 13)	84	72	88
With aerators only (run 3)	71	62	61
With both aerators and regional plants (run 16)	70	59	65

Solid Residuals Handling and Disposal

The next topic of interest is the handling of solid residuals within the regional model under different sets of exogenously imposed conditions. A summary of the key results is presented in Table 17 where we show, for each of the production runs, the quantities of solid residuals disposed of directly in landfills, the quantitites sent to incinerators in the region, and the quantities transported out of the region using the railhaul alternative. Also shown in this table are the total quantities of newsprint and corrugated containers collected in the region for reuse and the inputs of newsprint and corrugated to the proposed newsprint and linerboard plants that are considered in the regional model.

There are several interesting observations to be made on the basis of the output of the solid waste management-paper recycling module. First, the railhaul alternative is never used. It is simply too expensive in comparison with the other alternatives. Second, for all the combinations of conditions imposed on the model, only 7 out of the total of 23 municipal incinerators in the model are ever used. And when the air standards become too stringent--for example, level T--the municipal incinerators disappear entirely, as noted above.

Third, the input activity levels of the optional newsprint and linerboard plants indicate under what restrictions on ambient levels of air and water quality, and on landfills, it is economically feasible to recycle additional waste paper in the region. The linerboard plant is operating for most of the computer runs that we made, but at substantially different levels. Its lowest nonzero operating levels (around 100 tons/day input) are associated with T air quality standards and E or O water

Table 17

Results of Production Runs for Municipal Solid Residuals Disposal and Waste Paper Reuse

(tons/day)

	Computer Run							
	1	2	3	4	5	6	7	8
Characteristics of runs:								
air standards	T	T	E	E	T	T	E	E
water standards	T	E	T	E	T	E	T	E
landfill standards	H	H	H	H	M	M	M	M
in-stream aeration	+	+	+	+	+	+	+	+
regional sewage treatment	0	0	0	0	0	0	0	0
Management options for municipal solid residuals:								
Solid residuals to landfill	8,030	7,870	6,740	6,500	8,030	8,030	8,030	8,030
Solid residuals to incinerators[b]	0	0	1,280 (7)	1,370 (7)	0	0	0	0
Railhaul alternative	0	0	0	0	0	0	0	0
Paperstock collected for reuse:								
newsprint	149	308	153	298	141	141	143	146
corrugated	1,482	1,332	1,542	1,498	1,488	1,343	1,496	1,540
Paperstock inputs to new paper plants:								
newsprint plant	0.0	166	0.0	157	0	0	0.0	0.0
linerboard plant	248	98	318	263	253	109	261	305[d]

For footnotes, see page 87.

continued ...

Table 17 (continued)

(tons/day)

	Computer Run							
	9	10	11	12	13	14	15	16
Characteristics of runs:								
air standards	T	0	0	E	E	E	E	E
water standards	0	E	T	0	T	E	T	T
landfill standards	H	H	H	H	H	H	H	H
in-stream aeration	+	+	+	+	0	0	0	+
regional sewage treatment	0	0	0	0	0	0	+	+
Management options for municipal solid residuals:								
Solid residuals to landfill	8,020	4,960	5,120	6,020	6,340	6,520	6,530	6,780
Solid residuals to incinerators[b]	0	2,913 (7)	2,913 (7)	1,510 (7)	1,646 (7)	1,367 (7)	1,492 (7)	1,249 (7)
Railhaul alternative	0	0	0	0	0	0	0	0
Paperstock collected for reuse:								
newswprint	153	300	142	640	192	283	155	145
corrugated	1,235	1,514	1,539	1,542	1,542	1,540	1,504	1,537
Paperstock inputs to new paper plants:								
newsprint plant	0.0	158	0.0	499	0	142	0.0	0.0
linerboard plant	0	280	306[d]	307[d]	358[d]	305[d]	270	302

For footnotes, see page 87.

continued ...

Table 17 (continued)

(tons/day)

	Computer Run							
	17	18	19	23	25	26	Base[a]	Total available in the region
Characteristics of runs:								
air standards	∞	0	∞	0	0	0	0	
water standards	0	∞	∞	∞	0	0	0	
landfill standards	H	H	H	H	H	M	L	
in-stream aeration	0	0	0	+	0	0	0	
regional sewage treatment	0	0	0	0	0	0	0	
Management options for municipal solid residuals:								
Solid residuals to landfill	8,030	5,120	8,030	5,120				8,175
Solid residuals to incinerators[b]	0	2,910 (7)	0	2,910 (7)				9,340
Railhaul alternative	0	0	0	0				
Paperstock collected for reuse:								
newsprint	141	141	141	141	689	274	141	844[c]
corrugated	1,235	1,235	1,235	1,235	1,542	1,542	1,542	1,542[c]
Paperstock inputs to new paper plants:								
newsprint plant	0	0	0	0	548	133	0	
linerboard plant	0	0	0	0	308[d]	307[d]	307[d]	

For footnotes, see page 87.

continued ...

Table 17 (continued)

[a]"Base run" is a run of the Lower Delaware Valley Model made with low quality landfill (L), but with no marginal penalties on either gaseous or liquid discharges.

[b]The numbers in parentheses indicate the number of incinerators in the region with activities greater than zero.

[c]Taken from Table 8.

[d]Linerboard plant at its maximum capacity due to limit on the availability of used corrugated containers in the region.

quality standards--runs 2 and 6 (a rather strange result).[40] Its highest operating levels are associated with E and 0 air quality standards (runs 8, 11, 12, 13, 14, and the base run). It shuts down completely for the high effluent charge runs--for either liquid or gaseous discharges (runs 17, 18, and 19). The production level of the linerboard plant appears to be sensitive to the air quality standards, but insensitive both to the water quality standards (except when the effluent charges are very high) and to the costs of landfill (runs 25, 26, and the base run). The plant discharges its waterborne residuals to reach 5, a factor that partially explains its relative insensitivity to the water quality standards.

The newsprint plant is an entirely different story. First of all, the newsprint plant is extremely sensitive to the costs of landfill (runs 25, 26, and the base run). For conditions of high landfill costs and no restrictions on the discharges of liquid and gaseous residuals, the input of newsprint amounted to 548 tons per day. At this level of activity, used newspapers are collected from the single-family residences throughout the region, but mostly in Philadelphia. For medium landfill requirements, but still with no air or water quality standards imposed (run 26), the plant's optimal input level falls to 133 tons per day.

Second, the newsprint plant is quite sensitive to the water quality standards;[41] its activity level for the T water quality standards being consistently zero, or approximately so. This is explained by the plant's

---

[40] A linerboard plant capacity of less than 300 tons per day input would not be feasible. The costs in the model are based on a 300 to 500 tons per day capacity plant.

[41] The exception to this is run 9 where the activity level of the newsprint plant is zero and no water quality standards are imposed.

assumed location which is such that it discharges into reach 11, in the area of the worst water quality problems. At a water quality level of E, the activity level of the newsprint plant appears to be insensitive to the air quality standards. However, this is not true at a water quality level of 0. Then, inputs of newsprint decrease markedly as the air quality standards become more stringent. The newsprint plant does not operate at all for the cases of medium cost landfills (runs 5, 6, 7, and 8), except when no effluent charges are imposed on gaseous and liquid discharges (run 26), or for the very high liquid and gaseous effluent charge runs (17, 18, and 19).

Regional Environmental Quality

The last major set of results to be presented here are measures of the ambient quality predicted throughout the region for each of the production runs. First, in Table 18, we show the minimum, mean, and maximum concentrations of sulfur dioxide and suspended particulates over the 57 political jurisdictions, and the minimum, mean, and maximum concentrations of dissolved oxygen, fish, and algae in the 22 reaches of the eastuary. The jurisdictions, or reaches, experiencing the minimum and maximum concentrations are indicated in parentheses. In some cases, more than one jurisdiction (or reach) may be at the maximum value, although only one jurisdiction (or reach) is indicated; similarly for the minimum value.

Also shown in Table 18 are the minimum, mean, and maximum ambient concentrations that are predicted for the high quality landfill case with no air or water quality constraints imposed (run 25). The maximum concentration of sulfur dioxide for this situation is 290 $\mu gms/m^3$ in jurisdiction

Table 18

Distribution of Environmental Quality in the Lower Delaware Valley Region

		Computer Run						
		1	2	3	4	5	6	7
Characteristics of runs:								
air standards		T	T	E	E	T	T	E
water standards		T	E	T	E	T	E	T
landfill standards		H	H	H	H	M	M	M
in-stream aeration		+	+	+	+	+	+	+
regional sewage treatment		0	0	0	0	0	0	0
Sulfur dioxide	min	4 (39)	4 (39)	11 (43)	12 (43)	4 (39)	3 (39)	12 (43)
(μgms/m^3)	mean	22	22	53	54	23	22	54
	max	83 (16)	83 (16)	120 (16)	120 (16)	83 (16)	83 (16)	120 (16)
Suspended particu-	min	45 (3)	45 (3)	48 (3)	48 (3)	45 (3)	45 (3)	49 (3)
lates	mean	61	61	81	81	61	61	90
(μgms/m^3)	max	82 (15)	82 (15)	120 (17)	120 (24)	82 (15)	82 (15)	120 (24)
Dissolved oxygen	min	5.0 (17)	3.0 (17)	5.0 (17)	3.0 (17)	5.0 (17)	3.0 (17)	5.0 (17)
(mg/ℓ)	mean	6.2	5.4	6.3	5.4	6.0	5.0	6.2
	max	8.6 (1)	8.6 (1)	8.6 (1)	8.6 (1)	8.6 (1)	8.6 (1)	8.6 (1)
Algae	min	0.5 (20)	0.5 (20)	0.5 (20)	0.5 (20)	0.5 (20)	0.5 (20)	0.5 (20)
(mg/ℓ)	mean	0.9	0.9	1.0	1.0	0.9	0.9	1.0
	max	1.6 (10)	1.4 (10)	1.7 (10)	1.5 (10)	1.6 (10)	1.4 (10)	1.7 (10)
Fish	min	0.2 (20)	0.2 (20)	0.2 (20)	0.2 (20)	0.2 (20)	0.2 (20)	0.2 (20)
(mg/ℓ)	mean	0.5	0.5	0.5	0.5	0.5	0.5	0.5
	max	0.7 (10)	0.7 (10)	0.8 (10)	0.8 (10)	0.7 (10)	0.7 (10)	0.8 (10)

Note: Numbers in parentheses indicate the jurisdiction/RFF reach (as appropriate) with the minimum and maximum environmental quality indicator.

continued ...

Table 18 (continued)

		Computer Run						
		8	9	10	11	12	13	14
Characteristics of runs:								
air standards		E	T	O	O	E	E	E
water standards		E	O	E	T	O	T	E
land fill standards		M	H	H	H	H	H	H
in-stream aeration		+	+	+	+	+	O	O
regional sewage treatment		O	O	O	O	O	O	O
Sulfur dioxide ($\mu gms/m^3$)	min	12 (43)	3 (39)	19 (3)	19 (43)	12 (43)	12 (43)	12 (43)
	mean	54	22	130	126	54	54	54
	max	120 (16)	83 (16)	290 (16)	290 (16)	120 (16)	120 (16)	120 (16)
Suspended particulates ($\mu gms/m^3$)	min	48 (3)	45 (3)	50 (3)	50 (3)	48 (3)	49 (3)	48 (3)
	mean	80	61	99	99	81	81	81
	max	120 (17)	82 (15)	174 (17)	174 (17)	120 (17)	120 (24)	120 (17)
Dissolved oxygen (mg/ℓ)	min	3.0 (17)	0.9 (11)	3.0 (17)	5.0 (17)	0.8 (11)	4.9 (17)	3.0 (17)
	mean	5.0	3.3	5.3	6.3	3.3	6.3	5.2
	max	8.6 (1)	8.5 (1)	8.6 (1)	8.6 (1)	8.6 (1)	8.6 (1)	8.6 (1)
Algae (mg/ℓ)	min	0.5 (20)	0.6 (20)	0.5 (20)	0.5 (20)	0.6 (20)	0.3 (20)	0.5 (20)
	mean	1.0	1.1	0.9	1.0	1.2	0.9	1.0
	max	1.4 (10)	1.8 (15)	1.5 (10)	1.7 (10)	1.9 (15)	1.7 (10)	1.6 (10)
Fish (mg/ℓ)	min	0.2 (20)	0.3 (21)	0.2 (20)	0.2 (20)	0.4 (18)	0.1 (20)	0.2 (20)
	mean	0.5	0.5	0.5	0.5	0.5	0.4	0.5
	max	0.7 (10)	0.7 (2)	0.7 (10)	0.8 (10)	0.6 (2)	0.7 (10)	0.8 (10)

Note: Numbers in parentheses indicate the jurisdiction/RFF reach (as appropriate) with the minimum and maximum environmental quality indicator.

continued ...

Table 18 (continued)

		Computer Run						
		15	16	17	18	19	23	25
Characteristics of runs:								
air standards		E	E	∞	O	∞	O	O
water standards		T	T	O	∞	∞	∞	O
landfill standards		H	H	H	H	H	H	H
in-stream aeration		O	+	O	O	O	+	O
regional sewage treatment		+	+	O	O	O	O	O
Sulfur dioxide	min	12 (43)	12 (43)	3 (39)	18 (43)	3 (39)	18 (43)	18 (43)
(μgms/m^3)	mean	53	54	22	118	22	118	120
	max	120 (16)	120 (16)	83 (16)	290 (16)	83 (16)	290 (16)	290 (16)
Suspended particulates	min	49 (3)	48 (3)	45 (3)	50 (3)	45 (3)	50 (3)	50 (3)
(μgms/m^3)	mean	81	81	61	98	61	98	99
	max	120 (24)	119 (24)	82 (15)	174 (17)	82 (15)	174 (17)	175 (17)
Dissolved oxygen	min	5.0 (17)	5.0 (17)	1.2 (11)	5.2 (17)	5.2 (17)	6.0 (17)	0.9 (11)
(mg/ℓ)	mean	6.1	6.2	4.0	6.6	6.6	7.1	3.5
	max	8.6 (1)	8.6 (1)	8.6 (1)	8.6 (1)	8.6 (1)	8.6 (1)	8.6 (1)
Algae	min	0.3 (20)	0.5 (20)	0.6 (20)	0.2 (20)	0.2 (20)	0.2 (20)	0.6 (20)
(mg/ℓ)	mean	0.9	1.0	1.2	0.9	0.9	0.9	1.2
	max	1.7 (10)	1.7 (10)	1.6 (15)	1.7 (10)	1.7 (10)	1.7 (10)	1.8 (15)
Fish	min	0.1 (20)	0.2 (20)	0.3 (2)	0.0 (20)	0.0 (20)	0.0 (20)	0.4 (18)
(mg/ℓ)	mean	0.4	0.5	0.5	0.4	0.4	0.4	0.5
	max	0.7 (10)	0.7 (10)	0.6 (4)	0.7 (10)	0.7 (10)	0.7 (10)	0.6 (6)

Note: Numbers in parentheses indicate the jurisdiction/RFF reach (as appropriate) with the minimum and maximum environmental quality indicator.

16 (South Philadelphia);[42] for suspended particulates, the maximum concentration is 175 μgms/m^3 in jurisdiction 17.[43] The minimum concentration of dissolved oxygen in the estuary for this particular case is 0.9 mg/ℓ in reach 11; for algae, the maximum value is 1.8 mg/ℓ in reach 15; and for fish, the minimum value is 0.4 mg/ℓ in reach 18.

---

[42] The maximum sulfur dioxide concentration predicted by the RFF regional model is somewhat higher than the maximum predicted by EPA [41] (244 μgms/m^3; see Figure 3). This difference is due to the differences in emissions used in the two analyses (EPA employed their inventory of gaseous emissions, and RFF used outputs of the economic activity models embedded in the regional model). However, in making other comparisons between the two studies, differences in concentrations are due in part to the differences in emissions employed, but they may also be due partially to the stack aggregation procedures used in developing the air dispersion matrices incorporated in the regional model (discussed on pages 49 and 50). One aggregation procedure (which was used for all but two sources in the region), if not corrected, can produce systematic biases in the prediction of concentration levels. It causes the predictions to be higher than those obtained from an analysis where stack aggregation is not employed (assuming, of course, that percentage changes in discharges from all stacks in the group remain the same). The correction factor depends upon the number of stacks in the group, their stack characteristics, and their relative discharge rates, and in addition the correction factor varies for each receptor location. The other aggregation procedure (which was used for two major dischargers in the region) does not introduce biases into the prediction of concentration levels as compared with an analysis where stack aggregation is not employed (again, assuming that percentage changes in discharges from all stacks in the group remain the same). However, it should be noted that in cases where the stacks in a particular group have different stack characteristics and where percentage changes in discharges from the stacks in the group are allowed to differ, as they can in an analysis where stack aggregation is not employed, all stack aggregation procedures introduce errors, however small.

[43] The x-y coordinates selected to represent the air quality over political jurisdiction 17 unfortunately define a point in an industrial zone rather than in a residential area. The closest residential coordinates in the model are those of political jurisdictions 14, 16, 18, and 19. The closest residential area to the x-y coordinates representing political jurisdiction 17 is roughly 2,000 to 3,000 feet away.

Table 18 clearly indicates that for all situations explored, there is substantial variation in the quality of the air over the Lower Delaware Valley region and that portions of the region (most notably in Bucks and Montgomery Counties in Pennsylvania) have, for the most part, excellent air quality. On the water quality side, the highest dissolved oxygen levels in the estuary are always found in reach 1 near Trenton, N.J.; the lowest, when standards are imposed, in reach 17 near Wilmington, Delaware (see Figure 5).

Before going further with a discussion of the ambient quality aspects of the production runs, however, we should determine the best possible environmental quality attainable within the context of our model. We noted above that for both sulfur dioxide and suspended particulates the federal primary annual average standards (80 $\mu gms/m^3$ and 75 $\mu gms/m^3$, respectively) are exceeded in at least one political jurisdiction in the region for all runs of the model. (See in particular Table 11 and the results of run 17 in Table 18.) As stated before, we did not provide all the activities in the region with management options for reducing, and/or modifying, residuals discharges. Those that were not provided with management options were incorporated as background sources of residuals. In addition to this, many of the discharges from activities that are provided with management options (Table 6) cannot be reduced to zero (see Tables 15 and 16). Even though the model incorporates rather high removal efficiency options, in some cases the remaining discharges are substantial.[44]

---

[44]Most notably from the petroleum refineries in the regional model.

In Table 19, we indicate for the 57 political jurisdictions the annual average ambient ground level concentrations of sulfur dioxide due to (1) the uncontrolled sources by themselves (column 4) and (2) all sources, with maximum reduction in discharges from all managed sources in the model (column 5). The concentrations due to the uncontrolled sources represent the sum of discharges from point sources in EPA's inventory of gaseous emissions that were eliminated from the "over 25 $\mu gms/m^3$" category discussed above (column 2) and discharges from the 36 area sources in EPA's inventory that are located outside the Lower Delaware Valley region (column 3). Concentrations existing in spite of maximum reductions in discharges from all managed sources in the region were computed using the regional model with very high effluent charges ($200,000 per ton) on gaseous discharges (computer run 17).

We note from this table that all the concentrations due to the uncontrolled sources alone are below the federal primary standard for sulfur dioxide (80 $\mu gms/m^3$), the maximum concentration being 17.9 $\mu gms/m^3$ in political jurisdiction 15. However, when all the sources are considered in the model, one jurisdiction (in South Philadelphia) does not quite meet the primary standard: political jurisdiction 16 with a sulfur dioxide concentration of about 83 $\mu gms/m^3$.[45] The differences between the

---

[45]This result is consistent with measured levels of sulfur dioxide in the Metropolitan Philadelphia Air Quality Control Region during 1973 where an annual average concentration of 80 $\mu gms/m^3$ was recorded at one monitoring station in the region [58, Table A-4]. Unfortunately, the location of this station in the region is not known to the authors, but apparently it is not in Philadelphia where the highest sulfur dioxide levels are experienced. Of the 14 monitoring stations in the region reporting valid annual average sulfur dioxide concentrations in 1974, none are located in Philadelphia. But the one monitoring station in the region where both the

footnote continued

Table 19

Mean Annual Ground Level Concentrations of Sulfur Dioxide Resulting from Sources without Management Options and from All Sources with Maximum Reduction in Residuals Discharges

(μgms/m^3)

Political jurisdiction	Due to "under 25 μgms/m^3" point sources[a]	Due to 36 area sources outside the 11-county region[b]	Concentrations due to uncontrolled sources[c]	Concentrations due to maximum reduction in discharges from all sources[d]
1	3.3	1.5	4.8	16.5
2	4.9	1.6	6.5	14.1
3	0.6	3.5	4.1	6.2
4	3.1	2.3	5.4	11.5
5	2.5	1.3	3.8	7.4
6	1.1	1.2	2.3	4.2
7	0.9	1.6	2.5	4.5
8	3.8	1.2	5.0	22.0
9	3.4	1.2	4.6	31.8
10	2.7	1.2	3.9	24.3
11	2.2	1.2	3.4	27.4
12	1.8	1.3	3.1	14.6
13	2.5	1.2	3.7	11.0
14	10.8	1.2	12.0	55.7
15	16.7	1.2	17.9 (max)	41.9
16	14.2	1.2	15.4	83.4 (max)[e]
17	9.3	1.2	10.5	64.2[e]
18	7.0	1.2	8.2	35.1
19	10.4	1.2	11.6	50.0
20	6.7	1.2	7.9	23.7
21	4.3	1.3	5.6	17.6
22	8.8	1.3	10.1	28.4
23	11.3	1.3	12.6	37.8
24	14.0	1.4	15.4	28.2
25	9.3	1.5	10.8	19.0
26	14.0	1.4	15.4	24.8
27	14.9	1.3	16.2	33.6

For footnotes, see page 98. continued ...

Table 19 (continued)

Political jurisdiction	Due to "under 25 μgms/m3" point sources[a]	Due to 36 area sources outside the 11-county region[b]	Concentrations due to uncontrolled sources[c]	Concentrations due to maximum reduction in discharges from all sources[d]
28	10.5	1.3	11.8	24.2
29	5.0	1.3	6.3	16.9
30	10.5	1.4	11.9	22.5
31	5.7	1.4	7.1	15.4
32	5.4	1.6	7.0	12.4
33	5.5	1.8	7.3	11.8
34	3.2	1.3	4.5	10.5
35	3.3	1.4	4.7	10.4
36	1.6	1.4	3.0	5.4
37	5.3	1.5	6.8	11.6
38	2.7	1.6	4.3	8.3
39	0.6	1.3	1.9	3.1
40	4.5	2.1	6.6	10.4
41	3.1	2.8	5.9	8.9
42	2.1	2.1	4.2	6.7
43	0.6	1.9	2.5	3.7
44	6.1	3.4	9.5	12.7
45	2.3	4.4	6.7	9.5
46	1.7	5.2	6.9	8.3
47	4.0	1.5	5.5	12.0
48	3.8	1.9	5.7	9.8
49	1.7	2.0	3.7	7.4
50	11.2	1.2	12.4	39.8
51	5.6	1.3	6.9	20.2
52	5.4	1.2	6.6	26.8
53	2.6	1.2	3.8	16.2
54	1.0	1.3	2.3	6.3
55	1.7	1.2	2.9	21.6
56	1.4	1.2	2.6	8.7
57	1.0	1.6	2.6	8.1

For footnotes, see page 98.

Footnotes to Table 19:

[a]This group includes those activities in the IPP inventory of gaseous emissions that remain after all the activities listed in Table 6 have been removed. The ambient concentrations were computed using the IPP air dispersion model [44] with the calibration equations developed in an unpublished EPA study of the Philadelphia Metropolitan Air Quality Control Region [41].

[b]There are 36 area sources in EPA's inventory of gaseous emissions that are outside the Lower Delaware Valley region. The ambient concentrations due to these sources were computed using the IPP air dispersion model [44] with the appropriate calibration equations [41].

[c]This column represents the sum of the preceding two columns.

[d]Output of the Lower Delaware Valley regional model with no marginal penalties on liquid discharges (BOD, nitrogen, phosphorus, toxics, suspended solids, and heat) and with maximum marginal penalties on gaseous discharges ($200,000 per ton for sulfur dioxide and particulates). (Table 12, run 17.)

[e]Exceeds the federal primary annual average standard of 80.0 $\mu gms/m^3$.

concentrations due to the uncontrolled sources and those due to reduced discharges from all sources are substantial, if not surprising: approximately 68 μgms/m^3 for political jurisdiction 16 and roughly 54 μgms/m^3 for jurisdiction 17. Further, the model indicates that of the 68 μgms/m^3 difference for jurisdiction 16, 33 μgms/m^3 is attributable to one of the refineries in the regional model and 25 μgms/m^3 to another refinery in the model, both located in South Philadelphia on the east bank of the Schuylkill River.[46] Two other refineries across the estuary contribute smaller amounts of sulfur dioxide to political jurisdiction 16 in the range of 2 μgms/m^3 each. A similar situation holds for the observation of political jurisdiction 17.

Because the model predicts that the two Philadelphia refineries make rather large contributions to ground level concentrations of sulfur

---

24-hr and 3-hr standards were exceeded in 1974 is located in Philadelphia [59, Tables 3-5 and B-1]. This would imply that, although not reported, the highest annual average sulfur dioxide concentrations in the region are located in Philadelphia and that the maximum concentration reported for 1973 (80 μgms/m^3) is not the maximum in the region at all.

[46]The contributions to the annual average ground level concentration of sulfur dioxide in jurisdiction 16 reported here are based on an analysis using EPA's IPP air dispersion model, stack characteristics reported in EPA's inventory of gaseous emissions, and sulfur dioxide discharges from run 17 of the regional model. The discharges used in the analysis, 33.7 tons per day for Gulf Oil and 32.0 tons per day for Atlantic-Richfield Company (see Table 20), differ substantially from the reported discharges for 1974--Gulf Oil, 28 tons per day; Atlantic-Richfield (ARCO), 12 tons per day [60]. A comparison of various reported sulfur dioxide discharges for these two refineries is presented in Table 20.

The reported 1974 sulfur dioxide discharges for Gulf Oil and ARCO imply the following contributions to ground level concentrations in political jurisdiction 16: 21.0 μgms/m^3 and 12.2 μgms/m^3, respectively, for a combined total for the two refineries of 33.2 μgms/m^3. This combined contribution is substantially less than that predicted by the regional model (58 μgms/m^3) and would easily allow the regional model to meet the federal primary annual average sulfur dioxide standards in South Philadelphia (jurisdiction 16).

dioxide in political jurisdictions 16 and 17, and because it appears that these two sources comprise the major factor in the model's inability to meet the federal primary standard in South Philadelphia, even with very high effluent charges imposed ($200,000 per ton), it is worth commenting in more detail on their sulfur dioxide discharges and their reduction capabilities. Some of the key facts are set out in Table 20 where we see that the sulfur dioxide discharges from the two Philadelphia refineries in the model are substantial even under conditions of maximum reduction. It should also be noted, however, that the reported 1974 sulfur dioxide discharges from these two refineries are less, and in the case of Atlantic-Richfield substantially less, than those obtained as output of the regional model [60].[47]

The major reasons for the relatively low sulfur dioxide reduction efficiency achievable at the refineries in the model have already been mentioned. However, we add here for completeness some of the specific assumptions lying behind the removal efficiency limits obtained in the large refinery models on which the Lower Delaware Valley Model is based.

---

[47]For the Gulf Oil refinery, the difference between the reported sulfur dioxide discharge in 1974 (28 tons per day) and that obtained from the regional model (33.7 tons per day) is not large and is due primarily to the fact that the regional model was based on the 1970 sulfur dioxide discharge rate reported in EPA's inventory of gaseous emissions (85.6 tons per day). Had the regional model been based on the 1971 discharge rate (75.0 tons per day) reported elsewhere [60], it would have been possible to reduce Gulf Oil's sulfur dioxide discharge in the regional model to 29.5 tons per day.

The difference in the sulfur dioxide discharges reported for Atlantic-Richfield (ARCO) requires an explanation. ARCO began a five-year air quality improvement program in 1971. Coinciding with this program, the refinery has undergone a complete modernization, transforming one of the older refineries in the country into one with the latest in technical process and control equipment available today [60]. These modernization plans were not known to us when the original petroleum refinery models were constructed in the 1969-1972 period.

Table 20

Sulfur Dioxide Discharges and Reductions at Several Major Plants in the Metropolitan Philadelphia Region

(tons/day)

Plant Name	Reported Discharges			RFF Lower Delaware Valley Model				
	EPA's Inventory[a]	City of Philadelphia[b]		No Reduction (base case)		Maximum Reduction (run #17)		Maximum Removal Efficiency in Model (percent)
	1970	1971	1974	Discharge	Rank[c]	Discharge	Rank[c]	
Gulf Oil Corporation	85.6	75	28	84.2	13	33.7	1	60
Atlantic-Richfield Company	72.3	75	12	80.0	14	32.0	2	60
Eddystone Power Plant	190.6			190.1	1	0.03	43	99.98
Mercer Power Plant	162.2			162.2	2	0.13	33	99.92

[a]U.S. Environmental Protection Agency's Inventory of Gaseous Emissions for the Metropolitan Philadelphia Air Quality Control Region, 1970.

[b]William Reilly, Assistant Health Commissioner for Air Management Services, City of Philadelphia, as reported in "Control District News," Journal of the Air Pollution Control Association, Vol. 24, No. 12, December 1974, pp. 1202-1203.

[c]Ranked by magnitudes of sulfur dioxide discharges in the RFF regional model.

First, as previously noted, the use of natural gas for heating the refinery boilers is not allowed in the model, though the refinery can use desulfurized gas from its own process units. Second, no alternatives for removing sulfur dioxide in the stack gases are provided in the model. Third, the lowest sulfur content crude oil that can be charged is 0.4 percent, and in the catalyst regeneration unit in the cracker there is no way to reduce sulfur dioxide discharges except by lowering the sulfur content of the crude oil charged. Finally, in the process streams, there is leakage of sulfur dioxide in the removal of hydrogen sulfide and in the recovery of elemental sulfur from the H_2S. All this adds up to a maximum sulfur dioxide removal efficiency of 74 percent in the petroleum refinery models [13]. This removal efficiency, as we have noted, was reduced in the regional model by our adjustment of the base discharges.

For purposes of comparison, we also show in Table 20 the sulfur dioxide discharges and reduction capabilities of two of the larger electric power plants in the region--Eddystone and Mercer. For the case with no effluent charges imposed (base case), the sulfur dioxide discharges of these two plants rank 1 and 2 in the regional model, respectively. However, with very high effluent charges imposed (run 17), their discharges are reduced substantially and their sulfur dioxide discharge rank in the regional model drops to 43 and 33, respectively. In the last column of this table, we show, for purposes of comparison, the maximum sulfur dioxide reduction efficiencies for the power plants in the model. Note that these efficiencies are all substantially greater than those of the petroleum refineries. This is because we allow in the regional model the alternative of burning natural gas for generating electricity (Table B-2 of Appendix B).

In Table 21, we show for the 57 political jurisdictions, the annual average ground level concentrations of suspended particulates due to (1) the uncontrolled sources and (2) all sources, with maximum reduction in discharges taking place at the sources with reduction options. The construction of this table is similar to Table 19 with the following exceptions: particulate discharges in EPA's inventory of emissions from the area sources in the region, but not included in the home heat-commercial heat module, have been included as a background source (column 4). The concentrations from these sources are substantial--some of them in the range of 30 $\mu gms/m^3$--and it is perhaps unfortunate that management options were not provided in the regional model for this source category.

Considering only the contributions from the uncontrolled sources in the model, 4 political jurisdictions exceed (albeit only slightly) the federal primary standards for suspended particulates (75 $\mu gms/m^3$): jurisdictions 15, 23, 24, and 27 (all in Philadelphia). When all the sources in the regional model are considered, 9 political jurisdictions exceed the primary standards, the maximum concentration being 82 $\mu gms/m^3$ in political jurisdiction 15 (Center City Philadelphia).[48]

In Table 22, we indicate the concentrations of algae, fish, and dissolved oxygen that can be achieved in the Delaware Estuary with maximum

---

[48]This result is consistent with measured levels of suspended particulates in the Metropolitan Philadelphia Air Quality Control Region during 1974. Of the 39 monitoring stations in the region reporting annual average values of suspended particulates, 14 exceed the federal primary annual average standard of 75 $\mu gms/m^3$. Of the 14 stations where the primary standard was exceeded, 10 are located in Philadelphia. The highest (tentative) mean concentration recorded was 198 $\mu gms/m^3$ in Philadelphia, followed by 4 other stations in that city where the (tentative) mean concentrations recorded exceeded 100 $\mu gms/m^3$: 122, 120, 117, and 101 $\mu gms/m^3$ [59, Tables 3-5 and A-1].

Table 21

Mean Annual Ground Level Concentrations of Suspended Particulates Resulting from Sources Without Management Options and from All Sources with Maximum Reduction in Residuals Discharges

(μgms/m^3)

Political jurisdiction	Due to "under 25 μgms/m^3" point sources[a]	Due to 36 area sources outside 11-county region[b]	Due to discharges not included in home heat-commercial heat module[c]	Concentrations due to uncontrolled sources[d]	Concentrations due to maximum reduction in discharges from all sources[e]
1	1.7	4.9	6.2	47.8	49.4
2	1.8	5.1	5.4	47.3	48.6
3	0.6	6.5	2.2	44.3	44.7
4	1.3	5.6	3.7	45.6	46.4
5	4.2	5.0	6.3	50.5	51.7
6	2.5	4.9	5.7	48.1	49.1
7	1.9	5.3	4.4	46.6	47.3
8	3.1	4.4	17.4	59.9	62.5
9	2.6	4.5	15.8	57.9	61.2
10	2.7	4.5	13.5	55.7	58.7
11	2.2	4.5	10.5	52.2	55.8
12	1.9	4.7	8.4	50.0	51.9
13	4.9	4.6	11.4	55.9	58.0
14	5.7	4.3	27.1	72.1	76.0[f]
15	6.6	4.4	32.5	78.5 (max)[f]	82.0 (max)[f]
16	4.8	4.4	26.5	70.7	75.4[f]
17	5.4	4.4	22.9	67.7	73.3
18	3.5	4.4	21.4	64.3	67.6
19	3.9	4.4	27.5	70.8	74.8
20	3.6	4.4	20.2	63.2	65.9
21	3.5	4.5	18.1	61.1	63.6
22	4.2	4.4	28.7	72.3	75.5[f]
23	5.0	4.4	30.8	75.2[f]	78.6[f]
24	7.3	4.4	29.2	75.9[f]	79.2[f]
25	5.7	4.4	22.5	67.6	70.1
26	9.2	4.4	24.0	72.6	75.2[f]
27	6.3	4.4	32.3	78.0[f]	81.4[f]
28	4.4	4.4	25.3	69.1	72.0

For footnotes, see page 106.

continued ...

Table 21 (continued)

Political jurisdiction	Due to "under 25 µgms/m³" point sources[a]	Due to 36 area sources outside 11-county region[b]	Due to discharges not included in home heat-commercial heat module[c]	Concentrations due to uncontrolled sources[d]	Concentrations due to maximum reduction in discharges from all sources[e]
29	3.7	4.4	19.4	62.5	65.1
30	5.2	4.4	25.8	70.4	73.3
31	3.8	4.4	18.7	61.9	64.4
32	4.1	4.4	16.7	60.2	62.3
33	5.2	4.4	14.3	58.9	60.7
34	4.8	4.7	11.6	56.1	58.8
35	4.0	4.8	13.2	57.0	59.2
36	4.2	4.9	8.1	52.2	53.7
37	4.1	4.5	15.6	59.2	61.2
38	3.4	4.7	12.9	56.0	57.8
39	2.4	4.6	5.9	47.9	48.8
40	3.8	4.5	12.6	55.9	57.7
41	3.1	4.7	11.0	53.8	55.4
42	3.0	4.7	10.9	53.6	55.0
43	1.8	4.9	7.3	49.0	50.0
44	3.5	4.8	10.5	53.8	55.5
45	2.9	4.9	9.2	52.0	53.5
46	3.0	5.3	8.1	51.4	52.5
47	3.1	4.3	16.3	58.7	60.5
48	3.2	4.4	13.2	55.8	57.5
49	1.9	4.5	9.4	50.8	52.1
50	6.0	4.3	27.2	72.5	75.9[f]
51	3.4	4.3	20.5	63.2	65.5
52	3.1	4.3	20.2	62.6	65.2
53	2.0	4.4	11.1	52.5	54.4
54	1.1	4.4	5.7	46.2	47.2
55	1.4	4.6	8.6	49.6	51.7
56	1.2	4.6	6.5	47.3	48.4
57	0.9	5.1	4.1	45.1	46.1

For footnotes, see page 106.

Footnotes to Table 21:

[a]This group includes those activities in the IPP inventory of gaseous emissions that remain after all the activities listed in Table 6 have been removed. The ambient concentrations were computed using the IPP air dispersion model [44] with the calibration equations developed in an unpublished EPA study of the Philadelphia Metropolitan Air Quality Control Region [41]. The background concentration of 35.0 μgms/m^3 for suspended particulates is not included.

[b]There are 36 area sources in EPA's inventory of gaseous emissions that are outside the Lower Delaware Valley region. The ambient concentrations due to these sources were computed using the IPP air dispersion model [44] with the appropriate calibration equations [41]. The background concentration of 35.0 μgms/m^3 for suspended particulates is not included.

[c]Discharges in this group include those in EPA's area source inventory of gaseous emissions within the region, but not included in the home heat-commercial heat module. The method of calculating these discharges and the resulting ambient concentrations are presented in [61].

[d]This column represents the sum of the preceding three columns plus the calibration constant for suspended particulates of 35.0 μgms/m^3.

[e]Output of the Lower Delaware Valley regional model with no marginal penalties on liquid discharges (BOD, nitrogen, phosphorus, toxics, suspended solids, and heat) and with maximum marginal penalties on gaseous discharges ($200,000 per ton for sulfur dioxide and particulates). (Table 12, run 17.)

[f]Exceeds the federal primary annual average standard of 75.0 μgms/m^3.

Table 22

Concentrations of Algae, Fish, and Dissolved Oxygen in the Delaware Estuary for Maximum Reduction in Residuals Discharges[a]
-- With and Without Aerators --
(mg/ℓ)

Reach	With aerators[b]			Without aerators		
	Algae	Fish	Oxygen	Algae	Fish	Oxygen
1	0.95	0.51	8.6	0.95	0.51	8.6
2	1.17	0.53	8.5	1.17	0.53	8.5
3	1.18	0.55	8.1	1.18	0.55	8.0
4	1.16	0.55	7.7	1.16	0.55	7.4
5	1.11	0.53	7.5	1.11	0.53	7.1
6	1.12	0.53	7.6	1.12	0.53	7.2
7	1.01	0.49	7.2	1.02	0.49	6.7
8	1.16	0.52	7.2	1.17	0.52	6.7
9	1.38	0.61	7.5	1.38	0.61	7.0
10	1.66[c]	0.71	8.0	1.66[c]	0.71	7.6
11	1.18	0.54	6.3	1.19	0.54	5.7
12	1.38	0.57	6.9	1.38	0.57	6.3
13	1.13	0.51	6.3	1.14	0.50	5.6
14	1.00	0.44	6.2	1.01	0.43	5.5
15	0.89	0.36	6.3	0.90	0.36	5.6
16	0.72	0.28	6.4	0.73	0.28	5.7
17	0.47	0.19	6.0[d]	0.48	0.19	5.2[d]
18	0.28	0.09	6.5	0.29	0.09	5.8
19	0.26	0.06	6.9	0.27	0.06	6.5
20	0.20	0.04[d]	6.6	0.21	0.04[d]	6.4
21	0.34	0.05	6.6	0.34	0.05	6.4
22	0.60	0.15	6.9	0.61	0.16	6.7

[a]Output of the Lower Delaware Valley regional model with no marginal penalties on gaseous discharges (sulfur dioxide and particulates) and with maximum marginal penalties on liquid discharges ($200,000 per pound for BOD, nitrogen, phosphorus, toxics, and suspended solids; and $200,000 per million Btu's for heat). (Table 12, runs 18 and 23.)

[b]Aerators are at their maximum capacities for all reaches--1,000 horsepower per reach except for reaches 17 and 18 which have 1,500 horsepower aerators.

[c]Maximum levels. [d]Minimum levels.

reduction in wastewater discharges for two conditions: (1) with aerators and (2) without aerators. These results were obtained from the regional model with high effluent charges imposed on wastewater discharges--$200,000 per pound for BOD, nitrogen, phosphorus, toxics, and suspended solids, and $200,000 per million Btu for heat discharges. In contrast to the situation with the air quality of the region, high levels of water quality, as demonstrated by the dissolved oxygen levels, can be obtained all along the estuary. The minimum DO level was found in reach number 17: 6.0 mg/ℓ with aerators and 5.2 mg/ℓ without aerators. From this analysis, it appears there is very little question of the technical feasibility of achieving quite stiff water quality standards in the estuary. Whether such standards should be imposed is, of course, quite a different question.

VII. Special Features of the Results

In this section, we examine two features of the results: (1) the linkages among gaseous, liquid, and solid residuals, and among the various environmental media, as demonstrated by the output of the Lower Delaware Valley residuals management model; and (2) the distributional implications--for both costs and environmental quality--of alternative residuals-environmental quality management strategies for the 11-county Lower Delaware Valley region.

Linkages Among Forms of Residuals

One of the original objectives of applying to the Lower Delaware Valley region the residuals management modeling framework developed earlier at RFF was to investigate the linkages among gaseous, liquid, and solid residuals, and among the various environmental media, in a real world situation. This linkage question has two aspects that, for purposes of discussion, we wish to distinguish: (1) the evidence of linkages in integrated residuals management, and (2) the importance of considering these linkages in regional analyses.

The evidence of linkages among the forms of residuals and among the three environmental media may be obtained from a comparison of the output of the various computer runs using both aggregate costs, total and by category of discharger (Table 13), and acitivity levels, such as fuel usage, residuals discharges, residuals transport (solid wastes and sludge barging), and the collection and transportation of used paper in the region. Obtaining a measure of the importance of considering these linkages in regional analyses is more involved. Certainly one excellent measure of importance would be one based on the economic efficiency criterion: for example, the difference in the total discounted sum of capital, operating, and maintenance costs over some planning horizon of (1) an environmental quality management system designed and constructed on the basis of analyses that considered all three environmental media simultaneously, and (2) an environmental quality management system that was initially designed and constructed based on analyses of the three environmental media separately and then modified after initial construction to meet the air and water quality standards simultaneously. This difference in cost is a measure of the inefficiency implied by looking at each medium separately.

The results presented here demonstrate only that linkages are present; no efficiency measure of their importance is calculated. For this demonstration we will use four examples: three involve cost (objective function) information, and one involves the paper recycling activity levels.

The first example, involving total regional costs, demonstrates the linkages that exist between regional air and water quality. To illustrate, we use a portion of the matrix shown previously in Figure 7 (costs, in $ per day). When air quality standards are not considered in the analysis

Water Quality Standards	Air Quality Standards: 0	Air Quality Standards: E
0	$33,800 (#25)	$265,000 (#12)
E	$109,000 (#10)	$362,000 (#4)
T	$145,000 (#11)	$425,000 (#3)

(i.e., the 0 level), it costs the region $75,200 per day to go from existing conditions to the E level water quality standards. However, when E level air quality standards are considered in the analysis, the costs of improving water quality to the E level amount to $97,000 per day. Had there been no linkages at all between air and water quality, the increased costs would have been the same for both cases. In going from E level water quality to the T level, the differences are even more pronounced: $36,000 per day when air quality is not considered, rising to $63,000 per day when the E level air quality standard is considered in the analysis.

The second example also demonstrates the linkages that exist between regional air and water quality. This time we use the costs from the first module (MPSX 1)--the petroleum refineries, the steel mills, and the electric power plants. To illustrate, we use the information contained in the following matrix (cost data from Table 13, $ per day):

Water Quality Standards	Air Quality Standards: 0	Air Quality Standards: E
0	0 (#25)	$113,000 (#12)
E	$19,600 (#10)	$150,000 (#4)
T	$24,400 (#11)	$167,000 (#3)

When air quality standards are not considered in the analysis, the cost of achieving the set of E level water quality standards is $19,600 per day. When E level air quality standards are considered in the analysis, the costs of improving water quality jump to $37,000 per day, roughly a 90 percent increase. As with the total regional cost example above, the cost differences in going from E level to T level water quality standards are even more pronounced. When air quality is not considered in the analysis, the increased cost amounts to $4,800 per day. However, when E level air quality standards are considered simultaneously with water quality, the cost of improving the quality of the water from E level to T level rises sharply to $17,000, more than a threefold increase.

The third example involves the linkage between municipal landfill quality and air and water quality using the differences in the costs in runs 1 and 5, high and medium quality landfill, respectively, with T level standards for both air and water quality, as evidence of these linkages (Table 13 and Figure 9). If there had been no linkages between municipal

Total costs to the region:

Run 1 (high quality landfill)	$1,201,000
Run 5 (medium quality landfill)	1,164,000
Difference	$ 37,000

Solid residuals management costs (MPSX 5):

Run 1 (high quality landfill)	$ 78,700
Run 5 (medium quality landfill)	55,200
Difference	$ 23,500

landfill quality and air and water quality, the differences above would have been the same. The fact that they are substantially different indicates that linkages do exist. Obviously, there are substantial costs imposed on other sectors when higher quality municipal landfills are required. The linkages to these additional costs are through the air and water quality of the region.

The last example involves the output of the solid residuals management and paper recycling module (Table 17). Inputs of paperstock to the newsprint and linerboard plants for high (H) and medium (M) quality municipal landfills, and for various combinations of levels of air and water quality (O, E, T, and ∞), are shown in Figure 10. These inputs have been taken from Table 17 but have been rounded to the nearest 5 tons per day. This figure clearly indicates that the amount of recycling of used newsprint and corrugated containers in the region is extremely sensitive to the air and water quality standards imposed and to the municipal

High landfill quality (H)

Water quality standards \ Air quality standards	O	E	T	∞
O	550 (#25)	500 (#12)	0.0 (#9)	0 (#17)
E	160 (#10)	155 (#4)	165 (#2)	
T	0.0 (#11)	0.0 (#3)	0.0 (#1)	
∞	0 (#18)			0 (#19)

Medium landfill quality (M)

Water quality standards \ Air quality standards	O	E	T	∞
O	135 (#26)			
E		0.0 (#8)	0 (#6)	
T		0.0 (#7)	0 (#5)	
∞				

-- Newsprint Plant[a] --
(tons/day input)

High landfill quality (H)

Water quality standards \ Air quality standards	O	E	T	∞
O	310 (#25)	305 (#12)	0 (#9)	0 (#17)
E	280 (#10)	265 (#4)	100 (#2)	
T	305 (#11)	320 (#3)	250 (#1)	
∞	0.0 (#18)			0 (#19)

Medium landfill quality (M)

Water quality standards \ Air quality standards	O	E	T	∞
O	305 (#26)			
E		305 (#8)	110 (#6)	
T		260 (#7)	255 (#5)	
∞				

-- Linerboard Plant[a] --
(tons/day input)

Note: Numbers in parentheses indicate the production runs.

[a] Paperstock inputs taken from Table 17, rounded to the nearest 5 tons per day.

Figure 10. Inputs to the proposed newsprint and linerboard plants for various combinations of air and water quality standards imposed, and for high (H) and medium (M) landfill quality requirements

landfill quality requirements. It appears from this particular analysis that both air and water quality, along with landfill quality, would have to be considered simultaneously in order to determine the economic feasibility, and levels, of recycling used paper in the Philadelphia region.

In this section, we presented four examples illustrating the linkages among liquid, gaseous, and solid residuals, and among the ambient qualities of the various environmental media. Other examples of these linkages are presented elsewhere [11].

Distributional Implications

Two questions that we had hoped to be able to answer with the Lower Delaware Valley model are: (1) What does it cost to achieve various levels of environmental quality in the Lower Delaware Valley region? and (2) Who pays these costs and who benefits? The first question was discussed in detail above. The purpose of this section is to address the second question.

The outputs of the regional residuals management model that are relevant for a discussion of the "who pays and who benefits" issue are the distributions of environmental quality throughout the region and the distributions of the costs of improving the environmental quality of the region. It is generally asserted that Philadelphia now has, and will continue to have, the poorest quality environment in the region and that the residents of Philadelphia will have to bear the major share of the costs of improving the regional environmental quality. But is this true? And if so, how would these costs be passed along to the consumer, and how much will they amount to?

To shed some light on these questions and issues, we will examine, for various computer runs, the distributions of sulfur dioxide and suspended particulates concentrations over the region and the distributions of dissolved oxygen and algae along the estuary. In addition, we will examine the distributions of increases in certain consumer costs, among them: (1) percent increase in electricity bills, (2) percent increase in the costs of home heating, (3) percent increase in the costs of solid waste disposal, and (4) average increases in the costs of sewage disposal ($ per household per year).

Distribution of Environmental Quality. In Table C-2 of Appendix C, we present the distribution of sulfur dioxide concentrations over the Lower Delaware Valley region (all 57 political jurisdictions) for computer runs 1 and 3.[49] The maximum concentration of sulfur dioxide for both runs, as we have seen before (Table 18), is found in political jurisdiction 16 (South Philadelphia). In addition, of the 10 highest concentrations in Table C-2, 8 are located within Philadelphia (jurisdictions 14, 15, 16, 17, 18, 19, 23, and 27), 1 in Delaware County (jurisdiction 9), and 1 in Camden, New Jersey (jurisdiction 50).[50] Although not shown in this table, the suspended particulates distribution is similar to that of sulfur dioxide in that of the 10 highest concentrations, 9 are located within Philadelphia (jurisdictions 14, 15, 16, 19, 22, 23, 24, 26, and 27) and

---

[49] Each of these runs involves the higher level (T) of water quality and the high quality landfills (H). In both, the in-stream aeration, but not the regional treatment plant option, is included. Finally, in run 1 the model is asked to meet the stricter (T) air quality standards, while in run 3 the easier standards (E) are imposed.

[50] For computer run 1.

1 in Camden, New Jersey (jurisdiction 50).[51] For particulates, however, the maximum concentration for both runs 1 and 3 is found in jurisdiction 15 (in Center City Philadelphia). This certainly tends to confirm the assertion that the poorest air quality throughout the region will still be found in the Philadelphia-Camden area even after environmental quality controls become effective.

There is, however, another aspect of the distribution of air quality over the region that should be noted. In order to meet the T level sulfur dioxide standard (80 μgms/m^3) in South Philadelphia (run 1), dischargers throughout the entire region have had to reduce their sulfur dioxide discharges substantially. This is most clearly demonstrated in Table C-1 of Appendix C where the extra costs for home heating are at their maximum upper limits (indicating 100 percent conversion to natural gas) for practically all jurisdictions (run 1). For commercial heating (not shown in Table C-1), the situation is the same. As indicated in Table C-2, the maximum reductions were not required to meet sulfur dioxide standards in the jurisdictions in which the dischargers are located. In fact, the sulfur dioxide concentrations of all but one jurisdiction (16) are substantially less than the federal primary annual average standard of 80 μgms/m^3. This indicates that substantial cost increases (at least for home and commercial heating) are implied for all jurisdictions by the effort to meet the federal primary, or T level, sulfur dioxide standard in South Philadelphia. This represents a different type of linkage within the management model but nonetheless an important one.

---

[51]For computer run 1.

The situation with respect to water quality provides an interesting contrast with the air quality problem. To see this, consider the distributions of dissolved oxygen throughout the 22-reach estuary for computer runs 1, 2, 3, 13, 15, and 16 (Table 12) which are shown in Table C-3. (We use dissolved oxygen in this discussion as a surrogate for the distribution of estuary water quality.) When the E level water quality standards are imposed (run 2), the 8 poorest water quality reaches (which admittedly are of fairly high quality) are all located between reaches 11 and 19 (see Figure 5), from the confluence with the Schuylkill River to below Wilmington, Delaware. Using dissolved oxygen as an indicator of water quality, the estuary quality above the Schuylkill River is better than below. The critical, or poorest quality, reach is 17 near Wilmington. Thus, for this case, the lowest quality water is not found near Philadelphia but in the vicinity of Wilmington.

For the T level water quality standard (DO > 5.0 mg/ℓ), the distributions of the poorer quality reaches change (runs 1, 3, 13, 15, and 16) with the 8 poorest quality reaches being distributed between reaches 5 and 19, from above Philadelphia to below Wilmington, but with the identity of these reaches depending on whether or not the regional sewage treatment plants and/or in-stream aeration alternatives are employed. As before, the critical, or the poorest quality, reach is still 17 near Wilmington. And again, we note that the poorest quality water is not found near Philadelphia.[52]

---

[52]It should be noted, however, that the distributions of water quality along the Estuary presented here shed little light on the actual distributions of benefits and damages among the residents of the region. Nothing is said here about who uses the Estuary and for what purpose.

Distribution of Increased Consumer Costs. Now let us turn to an examination of the distribution of increased consumer costs for various residuals management strategies. In Table C-1 of Appendix C, we show the distribution of increased costs of electricity for computer runs 1 and 3, T level and E level air quality standards, respectively. For the stricter air quality standards, the largest increase in electricity costs, 15.8 percent, is found in New Castle County, Delaware (jurisdictions 1-4). The second highest increase, 13.6 percent, is found in Chester, Delaware, Philadelphia, Montgomery, and Buck Counties in Pennsylvania (jurisdictions 5-42). For the E level air quality standards, the situation reverses itself, and the Pennsylvania counties, including Philadelphia, are associated with the largest increase in costs (4.5 percent). Hence, as with the lowest water quality, the largest cost increases are not always associated with Philadelphia. Furthermore, in Camden, New Jersey, where the air quality is almost as poor as in Philadelphia, the increased electricity costs for the T level air quality standards amount to only 2.9 percent (run 1).

The distributions of the increased costs of both commercial heating and home heating are also revealing. (For home heating, see Table C-1. The distribution of increased costs of commercial heating are not reported in this paper.) As we mentioned before, in order to meet the strict (T level) air quality standards in South Philadelphia, virtually all jurisdictions convert to natural gas for both these uses. Thus, at least some of the costs of improving Philadelphia's air quality are imposed on other activities throughout the region. For the E level standards (run 3), however, complete conversion to natural gas in commercial heating--the highest cost alternative--occurs only in jurisdictions 16, 17 (both in South Philadelphia), and 18 (West Philadelphia). And for home heating, for the same air quality standard, conversion to natural gas does not occur at all. The maximum increase in home heating costs for this less restrictive case is associated with jurisdiction 23 located in North Philadelphia (30 percent increase).

Sewage disposal cost increases present another interesting example.[53] Wilmington (jurisdiction 2), Trenton (jurisdiction 44), and Camden (jurisdiction 50) are among the 10 jurisdictions exhibiting the highest average increased costs for sewage disposal ($ per household per year) for all runs. In fact, one of these three jurisdictions is always in the number one position--that is, absorbing the largest increase in costs of sewage disposal (Table 14). The remaining 7 highest cost jurisdictions are, however, located within the City of Philadelphia. When in-stream aeration is provided (runs 1 through 4), the highest cost increases within Philadelphia are associated with jurisdictions that are connected with the Philadelphia Southeast and Southwest sewage treatment plants.[54] (The highest increased

---

[53]The costs of sewage disposal have been allocated to the jurisdictions in the model on the basis of population (flow) served. The BOD concentrations of sewage have been calculated using different load factors for residential areas; mixed residential, commercial, and light industrial areas; and heavy industrial areas. The average increased costs per household for sewage disposal is computed on the basis of total sewage disposal costs divided by total housing units for each jurisdiction. Thus, all other factors the same, jurisdictions that are less than 100 percent sewered (Table B-3) will register smaller average increases than those that are 100 percent sewered. This accounts for the differences in increased costs for jurisdictions served by the same sewage treatment plant.

[54]For comparison purposes, of course, the costs of in-stream aeration should also be allocated among municipalities discharging wastewater to the Estuary. See footnote 37 on page 73.

costs for runs 1 and 2 are associated with Trenton and Camden, respectively (Tables 14 and C-1).) When no in-stream aeration is provided (run 13), however, the highest increased cost jurisdictions within Philadelphia shift to those that are connected to the Philadelphia Northeast sewage treatment plant. And in this case, a new jurisdiction outside Philadelphia (jurisdiction 11, in Delaware County) enters the list of the top 10. This particular jurisdiction is serviced by the following sewage treatment plants: Central Delaware County Authority, Chester, Eddystone, and Marcus Hook. The largest cost increase for sewage disposal in this case is found in Wilmington, Delaware (jurisdiction 2).

When regional sewage treatment plants (Table 7) are provided (run 15), the jurisdictions within Philadelphia with the largest cost increases shift back again to those connected to Philadelphia's Southeast and Southwest plants, and the costs to these jurisdictions actually increase over those found in run 13 where no regional options are provided. Note in Table 7 that both the Southeast plant and Southwest plant are eliminated in run 15 when a regional treatment plant is introduced. Thus, it is more costly for the Philadelphia jurisdictions to be connected to the regional treatment facility, even though total costs to the region for this option actually drop.[55]

When both in-stream aeration and regional treatment plants are provided (run 16), the largest cost increases within Philadelphia remain, as

---

[55]This reinforces the point made in footnote 32 on page 61. In comparing runs 13 and 15 (Table 13) where no in-stream aeration is provided, note that the total cost of municipal sewage treatment (MPSX 4) actually <u>increases</u> when regional treatment plants are introduced. The savings in costs accrue to the industrial dischargers (MPSX 6).

expected, with those jurisdictions connected to the Southeast and Southwest plants. In the latter two cases (runs 15 and 16), the largest increased costs occur in Wilmington (jurisdiction 2), the second highest in Camden (jurisdiction 50). For run 16, Trenton ranks third in increased sewage disposal costs.

Hence, here again, although the Philadelphia jurisdictions rank high in increased sewage disposal costs, they are not the highest in the region. One of the three cities--Wilmington, Trenton, and Camden--always exhibits greater average cost increases.

The distributions of the increased costs of disposing of municipal solid residuals are perhaps not as interesting as those of some of the other consumer costs. With a couple of exceptions, the distribution of costs among jurisdictions does not vary much. This is because for all runs made, all jurisdictions were required to maintain the same quality landfills, whether they be high quality (H), medium quality (M), or low quality (L). Where a significant variation did occur, it was associated with the high landfill quality runs (runs 1-4). Within this group of runs, the variations of increased costs among jurisdictions for the E level air quality standards (runs 3 and 4) were substantially greater than those associated with the T level standards (runs 1 and 2). This is because for the E level air quality standards, seven municipal incinerators (out of a total of 23), all available to Philadelphia, are used for solid residuals management purposes (see Table 17).[56] The result of this

---

[56]Four of these incinerators are located within Philadelphia. The other three--Chester, Darby, and Marple--are located in Delaware County. However, all seven incinerators are available to the City of Philadelphia.

is that the solid residuals disposal costs for the six jurisdictions within Philadelphia that use these incinerators are roughly 50 percent of those that do not have this opportunity.

For most computer runs, the largest increased costs of disposing of municipal solid residuals are associated with political jurisdictions within Philadelphia. It should be pointed out, however, that the minimum cost jurisdictions are also within Philadelphia. Where the economic feasibility of incinerators depends directly on the air quality standards (mostly within Philadelphia), the standards have an important impact on solid waste disposal costs for some jurisdictions (see runs 3 and 4).

The purpose of this section was to examine the distributions of environmental quality, and of the increased costs of improving regional environmental quality, among jurisdictions. We demonstrated that although the poorest air quality in the region would still be associated with Philadelphia, even after a cleanup campaign, this was not true of water quality. Nor will it necessarily be true that the costs associated with improving regional environmental quality will fall most heavily on that city. Indeed, an attempt to meet the federal primary air quality standards in Philadelphia promises to involve virtually the entire region in very large costs which would not be necessary to bring local air quality levels into line with the standards.

VIII. Policy Implications

Even from the limited number of computer runs which our research budget has allowed, and which have been discussed in some detail above, some tentative policy implications may be drawn.

1. The air quality over the Lower Delaware Valley region appears to be more of a problem than estuary water quality for at least two reasons: (a) in some areas of Philadelphia, it appears that it may be difficult, technologically, to achieve, and then maintain, the federal primary annual average standards for sulfur dioxide and suspended particulates (although, as we have noted above, there are indications that the sulfur dioxide primary standard is already being met throughout the region [59]);[57] and (b) it could cost the region up to six to eight times as much for air quality control (between $300 and $400 million per year) as for estuary water quality control (about $50 million per year). The lowest sulfur dioxide concentration at a location in South Philadelphia that we could achieve with our regional model was 83 μgms/m^3; for suspended particulates at another location, 82 μgms/m^3.[58] Had we not allowed complete conversion to natural gas by electric power plants, home and commercial heating, and the "over 25 μgms/m^3" industrial dischargers, the lowest attainable ambient sulfur dioxide concentrations would have been still higher.[59]

---

[57]The air quality violations in our regional model are so close to the federal primary standards, however, that we hesitate to make an issue of it here. The air dispersion model, and the residuals discharges, are simply not that accurate. Furthermore, as we have indicated above (footnote 46, p. 99), our regional model produces larger discharges of sulfur dioxide from two important sources in Philadelphia than those actually reported for 1974. The levels of air quality predicted by the model are, however, consistent with measured values in 1973 [58, Tables A-3 and A-4].

[58]The federal primary annual average standards for sulfur dioxide and suspended particulates are 80 μgms/m^3 and 75 μgms/m^3, respectively.

[59]As we pointed out above, however, the predicted levels of sulfur dioxide may be high due to (a) discrepancies in some of the discharge rates of major dischargers in the region, and (b) potential biases in the air dispersion matrices. See, in particular, the discussions in footnotes 42 (p. 93), 45 (p. 95), 46 (p. 99), 47 (p. 100), and 48 (p. 103).

2. The estuary water quality standards can be met in all cases, and there are various alternatives for meeting them: on-site sewage treatment, regional sewage treatment, and in-stream aeration. The regional alternatives appear to allow the region to meet the same ambient water quality standards more cheaply than it can by relying on individual treatment alternatives, but not all dischargers benefit to the same degree. Indeed, some dischargers actually pay more under the regional schemes than when management alternatives are restricted to local sewage treatment.

3. Under the conditions reflected in the model, roughly the 1970 situation, it appears that the railhaul alternative for disposing of solid residuals is too expensive. However, as landfill sites become scarce, this alternative will certainly become more attractive.

4. Again, under 1970 conditions, it appears that new municipal incinerators are not economically feasible alternatives for solid residuals management, especially with the stringent effluent controls on liquid and gaseous emissions that would be required.

5. Given a desire for improved air and water quality in the region, and the inevitable increase in landfill costs, increased paper reuse in the Lower Delaware Valley appears to be a promising management option.

6. From the set of cost analyses presented here, it appears that there are a variety of strategies for meeting the regional environmental quality standards that do not differ very much in total cost to the region. However, the distribution of levels of residuals treatment, of residuals discharges, and of costs among dischargers, all vary substantially from strategy to strategy. The costs do tend to fall on the major urban areas, though not on Philadelphia alone. At the same time, Philadelphia's air

quality problems are by far the most severe, even when every effort is made to meet the federal primary standards. This only emphasizes that there are two issues associated with improving the quality of the environment in the Lower Delaware Valley that must be addressed: (a) what level of quality is the region willing to pay for; and (b) who is going to pay for it. In general, given that there is a notable lack of mechanisms for interjurisdictional transfer payments in our system, these two issues cannot be separated. The chosen levels, and distribution of ambient quality, will reflect what patterns of initial cost distribution are possible given the regional structure and the features of its watercourses and atmosphere. This inseparability is recognized in the model discussed here, though the runs we are reporting do not include any in which the cost distributional constraints are varied.

Some Caveats

Before passing on to a discussion of the difficulties that we encountered in our illustrative case study, our assessment of research needs, and our own research plans, it will be worthwhile to pause and put the above policy conclusions in perspective by setting out in one place a list of some of the weaknesses of our regional management model. First, it should be noted that from the beginning this application was meant to be illustrative and not operational, though the model is based on actual data from the region and, in some areas, on the best data available to any person or group at the time of construction. In addition, the following specific caveats should be borne in mind:

1. The model was largely based on publicly available data of unknown quality and has only been partially calibrated. No field measurements of

any type have been made by the modeling team, and the implications of the results presented here have not been assessed in the field.

2. The model reflects conditions that existed in the region about 1970. For example, fuel costs in the model do not reflect recent increases due to worldwide changes in petroleum prices.

3. To keep the model as small as possible, all point sources (stacks) at the same x-y location in EPA's inventory of gaseous emissions were aggregated as noted before (p. 49). This aggregation procedure could have a substantial impact on the accuracy of predicted ground level concentrations, especially near large, multistack sources (see, in particular, the discussion in footnote 42 on page 93).

4. Some of the important industries in the region--particularly the petroleum refineries and the steel mills--have been represented in the model by "collapsed" versions of larger, generalized industry models. These collapsed versions have been calibrated to the production levels of specific plants, but other, perhaps more important, details peculiar to specific plants have not been included (see, in particular, the discussion in footnotes 46 and 47 on pages 99 and 100, respectively).

5. The receptor locations chosen to represent the air quality over the 57 political jurisdictions were not selected with respect to existing locations of residential areas or even to critical locations within each jurisdiction. They were simply chosen to be roughly in the central portion of each jurisdiction.

6. Certain categories of dischargers, such as the transportation sector, have not been incorporated in the model.

7. The costs reflected in the model are based on meeting constraints

(standards) on long-term average ambient concentrations (annual averages for air quality; low flow summer averages for estuary water quality), rather than on meeting shorter-term peak concentration constraints. The impact on costs and management strategies of the latter consideration is unknown to us.

8. The air quality predictions for the region are based on the Implementation Planning Program's (IPP) dispersion model, on 1970 meteorological data for Philadelphia, and on calibration relationships, all supplied by the U.S. Environmental Protection Agency. Clearly, our results depend heavily on these data and estimated relationships, but we have not investigated the sensitivity of these results to either alternative dispersion models or to alternative meteorological data.

9. The results presented here reflect certain policies in the region that may not be realistic. For example:

a. Complete conversion to natural gas in home and commercial heating activities is allowed in the regional model; it is also allowed in the electric power plants and some of the industrial dischargers.

b. Barging of municipal sewage sludge to sea is allowed for the larger cities along the estuary.

IX. Research Objectives and Lessons

As we pointed out in the Introduction, and again in the last section, this modeling application to the Lower Delaware Valley region was meant to be illustrative and not an operational study upon which to base public policy. Our primary interest in the effort was an application in the real world of methodology for analyzing alternative regional residuals management strategies that had been developed earlier at RFF [2,3,5]. Thus, even more important to us than the policy implications for the Lower Delaware Valley region presented in the last section are the results to date that have shed light on the three major research objectives stated at the beginning of the report:

1. to investigate the importance of including within a single analytical structure the linkages among gaseous, liquid, and solid residuals, and among the three environmental media;
2. to explore the feasibility of incorporating within a regional optimization model a complex, nonlinear aquatic ecosystem model; and
3. to explore ways of designing regional residuals management models to provide distributional information on, and to be able to constrain, costs and environmental quality such that these models would be useful in a legislative, as well as executive, setting. The _distribution_ of costs and environmental quality is often the central issue in regional environmental quality management, with efficiency considerations (such as least cost strategies) of secondary importance.

Intermedia Linkages and Tradeoffs

Ample evidence of the importance of linkages and tradeoffs among the ambient qualities of the three environmental media is provided by the results of the analyses, and four examples were presented in Section VII to demonstrate this. Although the examples do indicate the kinds, and extent, of the interdependencies among residuals forms and among environmental media, these examples may not convince the careful reader that it is really worth the effort to include these linkages in the analyses. To shed light on this issue, we must address two related questions: (1) how costly might it be to the region if these linkages are not considered explicitly in the analysis phase, and (2) how difficult and costly is it to include these linkages in the analysis of alternative regional residuals management strategies?

The first question is considerably more difficult for us to answer than the second. As we indicated above, one measure of the inefficiency (in terms of aggregate costs to the region) of not considering air, water, and landfill qualities simultaneously in the planning phase is the difference in the discounted sum of total costs over some planning horizon between:

(a) a regional system designed and constructed with all three environmental media considered, and

(b) a regional system designed and built with air, water, and landfill quality considerations kept separate, plus the costs of modifying this system later on to meet the air and water quality standards simultaneously.

Unfortunately, we have not made such an analysis since it would require restructuring the regional model to allow for "staged" runs, and our computer budget is already strained to the breaking point. Nonetheless, the results of the analyses do provide some, but perhaps less convincing, indication of the importance of considering these linkages. For example, while it appears that the costs involved in ignoring interactions would not be very large in comparison with the total costs to the region of meeting the

T (more stringent) level environmental quality standards, some are substantial on both an absolute and per capita basis. Perhaps even more important than the aggregate cost issue is the distributional one. If it turns out that residuals abatement facilities are actually underdesigned and that the ambient standards still cannot be met even after undertaking an extensive first phase abatement program--a likely situation if intermedia linkages are not considered in the analysis--resistance to additional abatement and costs will undoubtedly be experienced, and the distribution of the additional costs of abatement will certainly become a major issue. Attempts should be made to avoid such situations, if at all possible, and the inclusion of intermedia linkages is a first step in this direction.

Regarding the second question, we have more to report. Most important, our experience has been that the marginal costs of including the air-water-solids linkages has been modest--to the extent that we can separate the costs at all. Most of our early modeling problems involved data availability, but these were not worse for one medium than for another. Nor were they especially difficult for the linkage data themselves, such as the flyash generation rates in sludge incineration, or the suspended solids loads created in wet stack-gas scrubbers. In any case, recent legislation and regulatory activity has put something of a dent in all these data problems by forcing into the public domain much more information on costs and residuals discharges. The one drawback of the linked models, of course, is their size which has implications for the costs of analysis (to be discussed later).

All things considered, it appears to us that the benefit-cost ratio for linked models, especially where ambient environmental quality standards are at stake, would be very favorable to further development and application. Hence, we are inclined to argue for the practical significance of modeling these linkages.

Nonlinear Aquatic Ecosystems Models

We decided, early on in our research, to include in our regional model a relatively complex aquatic ecosystem model. We decided on this major research objective because, compared with the classical linear dissolved oxygen formulations, these newer models are potentially capable of providing, for public policy decisions, additional information on the resulting state of the natural environment, such as algal densities and fish biomass; and because they are alleged to provide more reliable predictions over wide ranges of river flows and residuals (wastewater) discharges. In addition, in the process of meeting ambient standards on dissolved oxygen, algal densities, and fish biomass, we were interested in examining the tradeoffs among discharges of various residuals, including organics (BOD), suspended solids, nitrogen, and phosphorus from the municipal treatment plants and industries, and the toxics and heat discharges from industrial and electric power plants. These discharge tradeoffs are difficult, and in some cases even impossible, to explore using classical water quality models.

The potential value of using aquatic ecosystem models in a resource management context is very great indeed. But there are substantial costs involved in incorporating them in a regional management model the way we did.[60] The inclusion within an optimization framework creates substantial

[60] For additional information, see [62].

computational problems and expense. In addition, at the current state-of-the-art, these models do not provide reliable predictions at trophic levels above algae, even for functional groups of species.[61] Whether or not it is worth the problems and expense of incorporating aquatic ecosystem models at this stage in their development depends on a number of factors and on the particular region being studied. For example, for analyzing alternative residuals management strategies in the Lower Delaware Valley, we do not feel the additional information provided is worth all the problems that are created by inclusion of these models. This conclusion is based principally on the fact that algae are not, at least currently, a major water quality problem in the Delaware Estuary.[62] In addition, this region is so large and complex from the point of view of the number of economic activities and the types and numbers of residuals discharges that any regional management model would be very large. And the larger the management model, the greater the problem that nonlinear aquatic ecosystem models pose.

On the other hand, since aquatic ecosystem models are generally considered reliable for predicting levels of dissolved oxygen and algae, they cannot be dismissed out of hand. In the context of a smaller region with many fewer discharges, the value of using these models may very well exceed the costs. This would certainly be true for a region where the environmental quality issue involved a tradeoff between algal blooms and an investment in tertiary (or advanced) wastewater treatment to reduce

---

[61]See the discussion by Dominic Di Toro in [63, pp. 345-352].

[62]Some researchers claim, however, that if the turbidity of the Delaware Estuary, caused by the suspended solids loading, were reduced, algae might become one of the major water quality problems there.

discharges of nutrients--nitrogen and phosphorus. The water quality problems of the Potomac River near Washington, D.C., and the controversy surrounding the investment in advanced wastewater treatment at the Blue Plains sewage treatment plant, provide an excellent example. The water quality problem in this part of the Potomac is algae, the region is small (with respect to the number of activities and dischargers), and the stakes are relatively high (1975 estimated capital costs of $587 million for the six major sewage treatment plants in the region, with $335 million of that amount representing the estimated costs of the additions at the Blue Plains plant) [64, Table II].

For the larger and more complex regions, a reasonable compromise at this time would be to use a linear dissolved oxygen model within the optimization framework. This would at least provide a relationship in the model between organic discharges and a measure of ambient water quality. Then, using the resulting wastewater discharge vector from the regional model as input to an ecosytem simulation model, the implications for other water quality measures, such as algal densities, could be investigated. This approach would take advantage of the additional information provided by aquatic ecosystem models, and at the same time would minimize the computational problems associated with the nonlinear optimization routine. Operating ecosystem models in the simulation mode is computationally straightforward. What we lose with this scheme, however, is our ability to meet ambient standards on algae at least cost, unless of course dissolved oxygen concentrations turn out to be a particularly good surrogate for algal densities. Unfortunately, for complex regions, with relatively large nutrient and industrial discharges, this is not likely to be the general case.

Another possibility would be to employ a linear phytoplankton model in the regional analysis. These models do exist, but they are not in as widespread use today as the nonlinear variety. In addition, certain restrictions must be placed upon their use in the analysis because of the assumptions made in their development. Such a model exists for the Delaware Estuary [65].

One major research implication of all this is the need for the development of efficient, large-scale, nonlinear programming algorithms that can deal with resource management problems of the type described here (to be discussed later on).

Distributional Information

Distributional information on both environmental quality and certain consumer costs is available as output of the Lower Delaware Valley regional model. This output was presented and discussed above in considerable detail and will not be repeated here. But because we feel that the generation of distributional information, especially on costs, is one of the most important features of the regional model presented here, we will take this opportunity to discuss the problems we had in trying to obtain certain kinds of distributional information that we desired.

Distributional information, especially on costs, can be used in two ways, depending on the analysis and on the type of management model employed: (1) it can be used in the unconstrained mode to provide information on the environmental and cost implications of alternative residuals management strategies; and (2) it can be used in the constrained mode to help shape the set of economically and politically feasible residuals management strategies that are selected for consideration. Both programming (optimization) and simulation (recursive) management models can be used for the former analyses, but only programming models are useful for the latter analyses. Since in most regions, the distribution of costs and environmental quality will be a more important issue than regional efficiency, we feel it is important to address the analytical problems associated with attempting to provide information based on constrained costs and environmental quality.

We have had relatively little difficulty using the Lower Delaware Valley regional model in constraining levels of regional environmental quality and generating information on the implied costs. And, of course, it would have been still easier, and certainly less expensive, to merely provide information on the implied costs and implied levels of regional environmental quality for various alternative residuals management strategies (however selected). But the real difficulty arose when we attempted to constrain both the levels of regional environmental quality and the distribution of costs simultaneously. There are two primary reasons for this difficulty:

1. when costs and levels of regional environmental quality are constrained simultaneously, infeasible solutions are commonplace (as one would expect a priori); and
2. the "real world" tradeoffs among distributions of costs, among levels of regional environmental quality and the environmental media, and between levels of environmental quality and costs, are extremely subtle and many and occur at the very top, and flattest portion, of the regional total cost response surface.

The first difficulty poses a problem for both linear and nonlinear programming formulations. (Simulations models, except in very simple situations, are of extremely limited use in this type of analysis.) If the computer run turns out to be infeasible, it may be obvious (from the dual values) which constraints need to be relaxed, but it is not at all obvious by how much these constraints should be relaxed. Clearly, we need much more operational experience here before we perfect this use of the regional model.

Regarding the second difficulty, current nonlinear programming algorithms are simply not practical for use with large regional applications, and may not be practical for the smaller (less complex) ones. Most nonlinear programming algorithms become less and less efficient as the optimum is approached. Thus, when the regional efficiency criterion is employed, it makes sense to stop these algorithms short of an optimum. Only modest cost savings (as a percentage of total regional costs) are at stake anyway. But in examining the tradeoffs among the distribution of costs and environmental quality, the important information is not only in the total regional cost dimension, but in a variety of other dimensions as well. It is the range of alternatives (or management strategies) for satisfying broadly stated societal objectives, and of course the resulting implications for individuals' costs and environmental quality, that become of major importance. And here is where the current crop of nonlinear programming algorithms let us down.[63] Unfortunately, for this kind of

---

[63]Regional environmental quality management models represent only one example of a large set of natural resource allocation problems that would benefit greatly from the development of nonlinear programming algorithms that could deal efficiently with practical, large-scale applications. We hope that these opportunities will one day be recognized by the applied mathematicians and operations researchers.

regional environmental quality analysis, our only choice at this point in time is to resort to linear programming techniques. And we are currently in the process of restructuring the Lower Delaware Valley regional model as an LP model by removing the nonlinear ecosystem model of the Delaware Estuary and replacing it with the Delaware River Basin Commission's linear dissolved oxygen model.

X. Difficulties Encountered and Research Needs

In the course of this regional modeling study, we have, as already indicated, run into a number of obstacles and difficulties which have cost us time and money, and which ultimately lie behind the caveats presented in the section on policy implications. For ease of exposition, we can group these difficulties into four headings:[64] (1) data availability, (2) model size, (3) nonlinearity, and (4) costs of analyses.

Difficulties Encountered

1. Data Availability. When we began this effort, data on specific residuals discharges were difficult to find in the public domain. And even if some idea of the discharges from a particular source could be ferreted out, there was only the sketchiest basis for the estimation of the costs of reducing those discharges. A large amount of time in the beginning, therefore, had to be spent in searching for such data as were available from the Delaware River Basin Commission (for wastewater dischargers).

---

[64] Some of the difficulties that we experienced, especially those associated with large-scale applications and with the use of a heuristic nonlinear programming algorithm, have been presented in considerable detail elsewhere [4].

An even larger investment of time was made in developing industrial models for petroleum refineries and steel mills in an attempt to get around the lack of cost information, while also reflecting the linkages among raw material input quality, production processes, product (output) specifications, and residuals generation to which we had been alerted by the work of Bower [16,17,66,67]. Unfortunately, the U.S. Environmental Protection Agency's inventory of gaseous emissions for the Metropolitan Philadelphia Air Quality Control Region, which turned out to be the single most important source of data as far as the regional modeling effort was concerned, was not available to us when we first started this application.

A similar kind of difficulty existed for models of the natural world. That is, actual observations of conditions in the water and atmosphere tied in some way to data on discharges were very scarce, and those data that existed were of doubtful quality (especially, as one would expect, the residuals discharges).[65] This meant that even if a model of the aquatic ecosystem, or of atmospheric dispersion, were to be constructed from first principles, published laboratory relationships, and so on, its calibration and verification would be difficult, and perhaps even impossible. And here we could not even begin to make good the lack of

---

[65]This is not meant to imply that no modeling efforts of the estuary, or of the atmosphere above the region, had ever been undertaken. On the contrary, the Delaware Estuary is one of the most well-studied bodies of water in the United States, the most well-known study of which is the Delaware Estuary Comprehensive Study (DECS) supported by the Federal Water Pollution Control Administration in the early 1960s [68]. But information on biological parameters and biological interrelationships was not readily available to us. Also, although information on gaseous discharges and an atmospheric dispersion model of the region eventually became available to us, this was not the case when we first started. Finally, 1970 Census information did not become available until the 1972-1973 period.

information by our own efforts. Data gathering of this sort is far too expensive in relation to the modest budget we were operating under.

Fortunately, things have begun to change. More and more discharge data are becoming available through the combined workings of our new environmental legislation and the Freedom of Information Act. Cost models are being constructed for many industries, especially under the RANN Program at the National Science Foundation.[66] In addition, the literature on treatment processes has grown enormously, especially under the sponsorship of the U.S. Environmental Protection Agency. The data describing actual ambient quality continue to expand, but there appears to be a need here for some common understanding of what data are to be sought and what methods and measures are to be used. There are indications, though, that even this situation is improving rapidly.

2. Model Size. In setting out to work on an integrated residuals management study of the Lower Delaware Valley region, we had only an imperfect understanding of how large the resulting model might eventually become. As we began to discover the potential size of such a model, we began to explore methods of reducing the size of the problem; for example, the aggregation of stacks, a reduction in the number of reaches in the estuary, the condensation of large linear programming models of industrial plants, and the elimination of sources below some arbitrary residuals discharge cutoff point. We have briefly discussed these methods here and plan to go into more detail in the final monograph on the project. Even

---

[66] For example, the RANN Program at NSF has been supporting the work of Russell Thompson at the University of Houston who has extended the RFF petroleum refinery model and constructed other such models. See, for example, [69,70,71].

after making these efforts, however, we still ended up with a very large regional model, the linear program portion of which had over 3,000 constraints in total, and about 8,000 columns, including about 800 residuals discharges. And this was not a sophisticated model with stochastic, or even integer variables. To summarize, we had four basic problems related to the size of the model: how to reduce the size; how to assess the loss of information implied by each reduction technique; how to work with, and obtain output from, the resulting model which was still very large; and how to interpret and present the flood of output which even a few production runs of such a model generate.

3. Nonlinearities. Perhaps our most serious difficulty in operating the model and in interpreting its output grew out of our inclusion of a nonlinear aquatic ecosystem model of the Delaware Estuary. Exploration of the use of such models, which were extremely new when this project began, was a high priority goal of the research. In order to pursue that goal, it was necessary for us to construct our own heuristic version of a nonlinear programming algorithm. The algorithm that we devised is based on the gradient method of nonlinear programming. This iterative procedure requires the computation of trial effluent charges at each step of the ascent process. The key to the efficiency of such a nonlinear program is the step-size selection routine, and we have never felt that the routines we devised were particularly good ones. In addition, of course, when dealing with a large, complex, nonlinear model, multiple optima are always a possibility. The combination of an apparently flat response surface, step-size selection problems, and our computation budget constraint encouraged us to arbitrarily stop all computer runs at around the 30th iteration.

However, we are sure that significant changes in the mix of discharge levels and the distribution of costs would have occurred had we let the process continue on. This procedure has hindered a comparison of various computer runs, especially a comparison of the distributional implications. As we mentioned previously, we are currently in the process of linearizing the regional model so that we can continue our investigations of the distributional aspects of regional environmental quality management.

4. Costs of Analysis. The mention of our budget constraints brings us to our last major problem--the costs of analysis. It is essentially impossible to calculate the cost of constructing the existing Lower Delaware Valley model, and since the effort has produced a number of independent outputs as joint products such a calculation would have little meaning in any case. But it is easy, if painful, to calculate the costs of a single computer run of the regional model. As we indicated above, this amounts to roughly $1,200 for a 30-iteration run of the nonlinear regional model under our current computer services setup. But even the newly constructed linear regional model will be expensive to operate. On experimental runs with only 114 shared constraints associated with the inclusion of the two air dispersion matrices (two 57 x 240 element matrices), the central processing unit (CPU) time approached 35 minutes. When the 90 rows associated with the linear dissolved oxygen model are added, the computer time will be greater still. So even the linear version of the regional model is not a panacea for all our computational problems. In short, it has been particularly frustrating for us to find that computational problems and expense stand in the way of much desirable exploration with the Lower Delaware Valley regional model.

Research Needs

The problems just discussed imply for us a set of high priority research needs, most of which are relevant not only to regional modeling but to any rational evaluation of a regional environmental policy proposal.

Data Needs. Data on existing levels of residuals discharges must become more widely available, cover a greater percentage of sources and of residuals types, indicate variability during the day, the week, the month, and the year, and reflect changes in exogenous conditions over long periods.[67] At the same time, data on actual ambient conditions ought to be gathered in such a way that calibration and verification of models of natural systems become possible. This will require both some hard thought about what data to gather, how often, and where, and considerable coordination in the data collection efforts. In this respect, it would be enormously helpful if government agencies would discourage isolated, specialized data gathering projects and actively encourage well thought out, comprehensive proposals dealing with a particular "environmental quality shed." Such an effort should include, among others, an analysis of the worth of data for various purposes and the design of a data collection network.

Modeling Needs. It is now almost certainly possible to construct cost of discharge reduction models for many important source types. It would be useful if such models, in some standardized format, were available

[67]Currently, such data, if they are gathered by sampling, are likely to amount to snapshots of conditions at one instant. If the data points represent estimates based on rules of thumb, as do much of the data in the residuals discharge inventories, they are at best long-term averages. In either case, there exists no measure of temporal variability.

"on the shelf" to prospective regional modeling teams. In addition, it seems to us that there would be a high payoff to the funding of careful studies of the complex industries currently beyond our capabilities--for example, the big petrochemical complexes.

On the other side of the coin, there is clearly room for considerable refinement of aquatic ecosystem models. As we pointed out above, the current generation of such models are generally conceded to be unreliable in predicting responses for trophic levels above phytoplankton, with the inaccuracies increasing rapidly as one moves from zooplankton to fish. Better data are, of course, central to improvement of these models, but it would also be useful if cooperation, and information exchanges, among modelers, laboratory scientists, and fisheries biologists in the field could be increased.

Operations Research Needs. As we emphasize above, it seems important to us that regional environmental quality management models include the linkages among disch rges of organic material, nutrients, and toxic materials, and various ambient levels of aquatic biota. But for this to be realized in practice, it will be necessary for nonlinear programming techniques to be significantly improved, since the necessary natural system models will certainly be nonlinear (with the exception of the linear phytoplankton models discussed above). Our heuristic nonlinear programming method is clearly too inefficient for use with large-scale models. What is needed badly is an efficient nonlinear programming algorithm that can deal effectively with these large-scale applications. The regional waste management problem provides an example of a class of problems in natural resource management that would benefit greatly from such development. Tradeoffs between water supply and water quality in the management of water resources systems, and between storage reservoirs and dike heights in flood management, represent two other examples where these algorithms would be useful. We hope that these opportunities will one day be recognized by the applied mathematicians and operations researchers, and that they will work to extend their admittedly impressive techniques in the direction of greater practical applicability. It seems likely that the fundamental problem here is one widely shared among academic disciplines--the professional rewards are for theoretical elegance and not for tough, applied work in which somebody else's lovely technique is made operational, and thus useful, in real decision situations.

XI. Further Research Plans and Final Comments

Further Research Plans

Our intentions at this point are to pursue two lines of inquiry related to the regional model we have been describing. Both require restructuring the Lower Delaware Valley regional model as an LP model by removing the nonlinear ecosystem model of the Delaware Estuary and replacing it with a linear dissolved oxygen model. (We are currently in the process of doing this.) This modification will allow us to investigate, more efficiently, the distributional implications of various policies. First, we shall examine: (1) the impact on regional efficiency of different constraints on the distribution of costs, given a set of ambient standards, and (2) how costs shift among groups of dischargers when different distributional constraints on costs are imposed. Second, we shall begin to explore the political side of the model in cooperation with Edwin Haefele.

That is, we shall examine how political "trades" on costs and on desired levels of environmental quality might be made among the local jurisdictions in such a way as to produce greater regional consensus on a desirable environmental policy for the region.

Final Comments

We have tried to be frank in this description of our regional environmental quality management model, its application to the Lower Delaware Valley region, and its several problems. Having set out to explore a number of modeling possibilities simultaneously, we found that some were easier to include than others. For example, bringing air, water, and solid waste problems into the regional model simultaneously is not particularly difficult or expensive per se. On the other hand, the inclusion of nonlinear aquatic ecosystem models brought us up against questions of the adequacy of ecosystem models themselves and of the capabilities of available nonlinear programming algorithms, and eventually forced us into developing a heuristic nonlinear programming algorithm to deal with large-scale applications such as the Lower Delaware Valley region. But it seems worthwhile to close by stressing that just because our experiment found large costs and dubious benefits in some directions, this is no reason to dismiss further work along these lines. Most importantly, it seems obvious that the coming generation of water quality decisions is going to involve nutrient and toxic discharges, their costs of removal, and their effects on stream quality. These questions (and enormously expensive their implications will be) cannot be addressed in any but the most casual and ad hoc ways without the use of water quality, and ecosystem, models which make reasonably accurate predictions at least of the effects of discharge reduction on algal concentrations. To provide this information, it will probably require models that include functional groups up to zooplankton. Public policy would, however, be immensely better informed if we could relate discharges to higher trophic levels, particularly carnivorous fish, as well. It may be that such models will only be useful in a simulation context for the foreseeable future (that is, least cost ways of meeting ambient water quality and biological standards will be beyond the management model's capabilities). Even this, however, will give society some better grasp of its options and their costs.

Finally, it is clear from our analyses that ambient environmental quality standards can be met through varying combinations of strategies which can imply substantially different distributions of costs among the public and private sectors, among the residents of the various subregions, and among different income groups. It is also clear that at the local level, at least, the distribution of the costs of improving and/or maintaining environmental quality will be the central issue in determining the political feasibility of different strategies, with total regional costs of secondary importance. Thus, to the extent possible, information on cost distributions of each strategy should be generated and presented, and constraints on the distribution of costs should be used to formulate technologically, economically, and politically feasible regional residuals management strategies for consideration.

APPENDIX A

Mathematical Statement of the Regional Residuals Management Model

The regional residuals-environmental quality management model may be stated formally as

$$\min \left\{ F_1(X,S) = C \cdot X_1 + P(X_2,S) \right\} \tag{1}$$

$$\text{s.t.} \qquad AX \geq B \tag{2}$$

$$X \geq 0; \tag{3}$$

where $P(X_2,S)$ represents the total regional penalties associated with exceeding the ambient environmental quality standards, expressed functionally as

$$P(X_2,S) = \sum_{i=1}^{q} p_i\Big(S_i, R_i = h_i(X_2)\Big); \tag{4}$$

and where $F_1(X,S)$ is a nonlinear regional objective function; X is a vector of activity levels partitioned into two parts: X_2, residuals discharges, and X_1, all other activities; S is a vector of ambient environmental quality standards; $AX \geq B$ is a set of linear equality and inequality constraints, including constraints on minimum production levels and on regional requirements for space heat, electricity, and the disposal of municipal liquid and solid residuals; C is a vector of linear cost coefficients representing the costs of production, and the handling, modification, and disposal of residuals; R is a vector of ambient environmental quality with $R = h(X_2)$ representing the environmental models which relate residuals discharges, X_2, with ambient quality; and finally, $p_i(S_i,R_i)$, $i = 1,\ldots,q$, are the individual penalty functions associated with exceeding the environmental standards, S_i, $i = 1,\ldots,q$. For reasons that we will see below, we require that the function $P(X_2,S)$ be continuous and have continuous first derivatives.

Because the constraint set, $AX \geq B$, is linear, we are able to use a standard linear programming code[1] to select better and better positions along the nonlinear response surface, Eq. (1). For this purpose, we linearize the objective function, $F_1(X,S)$, at a point $X^{(k)}$, and constrain the allowable distance of travel to the next point, $X^{(k+1)}$. Establishing a new set of bounds for these constraints at each step in the procedure is analogous to selecting a step size in other gradient methods of nonlinear programming. The efficiency of the optimization scheme depends directly on the selection of these bounds.[2] The modified linear problem may now be stated, for the (k+1)th step (iteration), as

$$\min \left\{ F_2^{(k+1)} = C \cdot X_1^{(k+1)} + \frac{\partial P(X_2^{(k)},S)}{\partial X_2} \cdot X_2^{(k+1)} + \gamma \right\} \quad (5)$$

$$\text{s.t.} \qquad AX \geq B \quad (2)$$

$$X \geq 0 \quad (3)$$

$$X_2^{(k+1)} \leq \beta^{(k+1)} \quad (6)$$

$$X_2^{(k+1)} \geq \alpha^{(k+1)} \quad (7)$$

where β and α are, respectively, upper and lower bounds on the residuals discharge variables at the (k+1)th iteration; C is a vector of linear

---

[1]We use IBM's linear programming code MPSX.

[2]For information on the selection of these bounds, see [2,4].

cost coefficients; $\partial P(X_2,S)/\partial X_2$ is a vector of marginal penalties; and γ is the intercept.

To employ this method, we must be able to evaluate the total penalties, $P(X_2,S)$, for various sets of residuals discharges. This requires solving the relevant environmental models for a given discharge vector, X_2, to determine the resulting state of the natural world, and then comparing this state with the ambient environmental quality standards. In addition, the vector of marginal penalties, $\partial P(X_2,S)/\partial X_2$, must be evaluated for each state of the natural world which has been computed. This analysis requires a side computation involving the environmental models, and this computation must be made at each step in the iterative procedure.[3]

At the optimum, the marginal penalties, $\partial P(X_2,S)/\partial X_2$, are in fact the optimal set of effluent charges. Because the optimal set of both residuals discharges, X_2, and effluent charges are output of the analysis, either effluent standards or effluent charges could be used as the policy instrument for ensuring an efficient, or "optimal," management strategy.

---

[3]For more detail, see [5,7].

Appendix B

Activities in the Lower Delaware Valley Model, Their Locations, and Their Management Options

Contents

Table B-1

Residuals Generation and Discharge Activities in the Lower Delaware Valley Model
-- Part I: Industrial Plants (Module 1) --

Plant	Location						
	City	County	State	Coordinates		Estuary reach[a]	
				x	y	RFF	DRBC
Petroleum refineries:							
Getty Oil Corp	Delaware City	New Castle	DE	446.0	4,381.7	19	26
Mobil Oil Corp	Paulsboro	Gloucester	NJ	477.7	4,410.2	12	16
Texaco	Westville	"	NJ	486.6	4,413.2	11	15
Arco	Philadelphia	Philadelphia	PA	482.6	4,420.6	11	15
British Petroleum	Trainer	Delaware	PA	465.0	4,407.1	14	18
Gulf	Philadelphia	Philadelphia	PA	482.7	4,419.1	11	15
Sun Oil Co	Marcus Hook	Delaware	PA	463.8	4,406.9	15	19
Steel mills:							
Fairless	Fairless Hills	Bucks	PA	521.9	4,445.0	2	2
Alan Wood	Conshohocken	Montgomery	PA	473.1	4,437.5		
Luckens	Coatesville	Chester	PA	429.7	4,425.5		
Phoenix (Phoenixville)	Phoenixville	"	PA	456.1	4,442.7		
Phoenix (Claymont)	Claymont	New Castle	DE	462.0	4,406.4	15	19
Power plants:							
Richmond	Philadelphia	Philadelphia	PA	493.6	4,425.7	8	10
Delaware	"	"	PA	489.1	4,423.9	9	12
Schuylkill	"	"	PA	483.8	4,421.2	11	15
Southwark	"	"	PA	488.1	4,418.0	10	13
Eddystone	Eddystone	Delaware	PA	472.4	4,412.2	13	17
Chester	Chester	"	PA	466.9	4,408.6	14	18
Barbados		Montgomery	PA	469.3	4,440.2	11	15
Cromby		Chester	PA	454.8	4,444.5	11	15
Burlington		Burlington	NJ	510.9	4,435.9	5	5
Mercer		Mercer	NJ	522.8	4,447.3	1	1
Quakertown		Bucks	PA	471.1	4,476.3		
Lansdale		Montgomery	PA	475.6	4,454.8		
Penn RR	Philadelphia	Philadelphia	PA	484.2	4,423.0	11	15
Deepwater		Salem	NJ	456.4	4,392.9	17	22
Greenwich		Gloucester	NJ	475.1	4,408.5	13	17
Edgemoor		New Castle	DE	456.8	4,399.3	17	21
Delaware City	Delaware City	New Castle	DE	445.8	4,382.6	19	26

[a]Resources for the Future and Delaware River Basin Commission reach numbering schemes, respectively.

Table B-1 (continued)

-- Part II: Large Dischargers of Gaseous Residuals (Module 3) --

(Taken from the U.S. Environmental Protection Agency's Inventory of Gaseous Emissions, listed in order of contribution to ground level concentrations of sulfur dioxide--Plants 1 to 52--and suspended particulates--Plants 53 to 75.)

No.	Name of Plant	County	State	Location				Rank			
				Coordinates		Estuary Reach[a]		Based on Maximum Ground Level Concentration		Based on Discharge	
				x	y	RFF	DRBC	SO_2	Suspended Particulates	SO_2	Particulates
1	Olin Chemical	Gloucester	NJ	480.1	4410.4			5	332	24	331
2	National Foam	Chester	PA	449.1	4423.3			7	33	170	225
3	Rohm & Haas[b,c]	Bucks	PA	511.3	4437.9	4	4	8	7	27	31
4	General Electric	Montgomery	PA	465.9	4437.6			10	83	119	203
5	Rohm & Haas[c]	Philadelphia	PA	494.7	4428.2	7	8	11	62	28	44
6	Boeing	Delaware	PA	473.9	4412.3			12	99	66	146
7	McGuire A.F.B.	Burlington	NJ	534.6	4431.1			13	4	48	11
8	Allied Chemical[b]	New Castle	DE	462.7	4406.0			14	273	20	158
9	Wyeth Laboratories	Chester	PA	449.5	4423.1			15	104	113	224
10	Hercules Powder[b]	Burlington	NJ	514.3	4437.8			16	81	42	88
11	Philadelphia Naval Base[b]	Philadelphia	PA	484.1	4415.4			17	78	68	126
12	Stauffer Hoechst	New Castle	DE	444.4	4382.2			18	335	19	335
13	University of Delaware	New Castle	DE	435.6	4392.3			19	97	49	105
14	New Jersey Zinc[b,c]	Camden	NJ	488.8	4415.0	10	14	20	66	38	84
15	Atlas Chemical[b]	New Castle	DE	454.0	4393.4			23	125	54	133
16	Continental Oil	Delaware	PA	465.3	4408.1			24	185	46	317
17	Hercules Powder	New Castle	DE	445.6	4401.0			26	101	121	150

For footnotes, see page 156.

continued ...

Table B-1, Part II (continued)

No.	Name of Plant	County	State	Location: Coordinates x	Location: Coordinates y	Location: Estuary Reach[a] RFF	Location: Estuary Reach[a] DRBC	Rank: Based on Maximum Ground Level Concentration SO_2	Rank: Based on Maximum Ground Level Concentration Suspended Particulates	Rank: Based on Discharge SO_2	Rank: Based on Discharge Particulates
18	Penn Central Trans	New Castle	DE	455.4	4399.5			27	124	120	220
19	Weyerhauser	Montgomery	PA	477.4	4434.0			28	136	45	125
20	Dupont-Carney Pt.[b]	Salem	NJ	458.0	4395.7			30	120	53	117
21	Firestone	Montgomery	PA	447.8	4453.4			31	39	44	40
22	Delaware State Hospital	New Castle	DE	450.7	4394.5			32	132	123	223
23	Valley Forge Hospital	Chester	PA	454.0	4441.3			33	133	100	193
24	Wyeth Laboratories	Chester	PA	455.3	4432.5			34	130	142	245
25	Heintz	Philadelphia	PA	489.7	4431.6			36	141	91	182
26	Milprint	Chester	PA	440.9	4428.3			40	145	181	279
27	Pillsbury	Montgomery	PA	456.7	4472.8			42	143	212	319
28	Dupont-Edgemoor[b,c]	New Castle	DE	457.2	4400.0	17	21	43	48	29	51
29	Burroughs	Chester	PA	459.4	4432.7			44	127	224	318
30	Willow Grove Naval Air Station	Montgomery	PA	480.0	4449.6			46	30	96	49
31	Leeds & Northrup	Montgomery	PA	475.7	4451.3			47	156	124	217
32	Bestwall[d]	Camden	NJ	494.4	4425.2			48	109	76	118
33	Westinghouse	Montgomery	PA	475.4	4412.6			49	43	43	89
34	Dupont-Experiment Lab.	New Castle	DE	451.0	4402.7			50	107	56	107
35	N.J. State Hospital	Camden	NJ	494.6	4404.1			51	152	82	168
36	SunOlin	New Castle	DE	462.8	4406.3			53	87	23	28
37	Koppers	New Castle	DE	447.0	4396.0			55	208	135	276

For footnotes, see page 156.

continued ...

Table B-1, Part II (continued)

No.	Name of Plant	County	State	Location				Rank			
				Coordinates		Estuary Reach[a]		Based on Maximum Ground Level Concentration		Based on Discharge	
				x	y	RFF	DRBC	SO_2	Suspended Particulates	SO_2	Particulates
38	Philadelphia Federal Buildings	Philadelphia	PA	485.0	4429.6			56	170	150	266
39	Philco	Montgomery	PA	476.2	4453.3			57	177	108	216
40	Shell Chemical[b,c]	Gloucester	NJ	486.3	4408.8	11	15	58	98	178	144
41	Dave Corporation	Chester	PA	439.9	4429.3			59	163	101	194
42	Dupont-Chambers[b,c]	Salem	NJ	456.7	4393.5	17	22	61	123	37	65
43	RCA	Camden	NJ	497.0	4418.7			62	114	177	212
44	FMC[b]	Delaware	PA	464.7	4407.7			63	202	22	83
45	Chamberlain[b]	Burlington	NJ	510.4	4436.1			64	118	87	138
46	Westfield Paper	Montgomery	PA	472.4	4435.5			67	173	115	202
47	Certainteed St. Gobain	Montgomery	PA	480.9	4444.5			68	174	118	215
48	Atlas Chemical	New Castle	DE	453.2	4404.5			69	187	140	247
49	Chrysler	New Castle	DE	435.0	4390.7			70	139	95	159
50	Oxford Royal Mushrooms	Chester	PA	424.8	4406.8			71	167	217	326
51	Dana Corporation	Montgomery	PA	446.1	4454.2			72	178	171	227
52	Congoleum Industries	Delaware	PA	464.6	4408.2			73	214	80	163
53	Crown Products	Philadelphia	PA	492.2	4427.7			301	3	298	25
54	Continental Distilling	Philadelphia	PA	487.5	4418.9			314	5	311	54
55	Stauffer Chemical[c]	Bucks	PA	519.8	4449.1	1	1	147	6	282	113
56	Holmsburg Prison	Philadelphia	PA	498.4	4431.7			83	8	72	7
57	Lavino[c]	Philadelphia	PA	490.4	4424.8	9	11	306	10	303	36

For footnotes, see page 156.

continued ...

Table B-1, Part II (continued)

No.	Name of Plant	County	State	Location				Rank			
				Coordinates		Estuary Reach[a]		Based on Maximum Ground Level CConcentration		Based on Discharge	
				x	y	RFF	DRBC	SO_2	Suspended Particulates	SO_2	Particulates
58	Wenczel Tile[c]	Mercer	NJ	522.3	4454.1	e	e	197	12	281	71
59	Shieldalloy	Gloucester	NJ	498.1	4376.9			275	13	261	111
60	Flintkote[c]	Camden	NJ	489.0	4420.5	9	12	311	14	308	73
61	Penn Packing; George Sall Metals[c]	Philadelphia	PA	491.9	4427.4	8	10	95	16	183	53
62	Hoegannes[c]	Burlington	NJ	501.2	4430.1	7	7	287	20	284	66
63	U.S. Gypsum[c]	Philadelphia	PA	481.9	4419.7	11	15	317	22	314	60
64	General Refractories[c]	Philadelphia	PA	494.0	4426.3	8	9	298	24	295	79
65	Penn Galvanizing[c]	Philadelphia	PA	491.4	4427.0	9	11	304	27	301	153
66	Cattie Galvanizing[c]	Philadelphia	PA	489.6	4425.1	9	11	307	28	304	154
67	Boyles Galvanizing[c]	Philadelphia	PA	489.6	4425.3	9	11	308	29	305	155
68	Reed; Ruddle[c,d]	Salem	NJ	456.7	4391.9	18	23	325	31	323	115
69	Kind & Knox Gelatin	Camden	NJ	489.7	4422.6			167	34	143	140
70	National Lead[c]	Philadelphia	PA	489.8	4425.3	9	11	181	35	146	29
71	Eastern St. Hospital	Philadelphia	PA	503.8	4443.8			170	36	219	67
72	Franklin Smelting[c]	Philadelphia	PA	492.3	4426.3	9	11	300	41	297	119
73	Philadelphia Coke	Philadelphia	PA	494.8	4427.4			141	42	64	27
74	Avisun	New Castle	DE	448.8	4388.7			121	215	35	92
75	Fort Dix	Burlington	NJ	532.9	4429.2			214	260	269	280

For footnotes, see page 156.

Footnotes to Table B-1, Part II:

[a]DRBC designates the Delaware River Basin Commission's reach numbering scheme; RFF, Resources for the Future's reach numbering scheme.

[b]The liquid residuals discharges from, and liquid residuals management options for, these 12 plants are included in Module 6.

[c]These 20 plants have suspended solids (liquid) discharges in this module from wet scrubbers used to remove particulates in the stack gases.

[d]Taken from USEPA's inventory of gaseous emissions, but the name of the plant could not be verified.

[e]Does not discharge to the Delaware Estuary.

Table B-1 (continued)

-- Part III: Municipal Sewage Treatment Plants (Module 4) --

No.	Name of plant	Location				
		City	County	State	Estuary reach	
					RFF	DRBC
1	Trenton	Trenton	Mercer	NJ	1	1
2	Morrisville	Morrisville	Bucks	PA	1	1
3	Bordentown	Bordentown	Burlington	NJ	2	2
4	Lower Bucks	Levitt Town	Bucks	PA	3	3
5	Florence	Florence	Burlington	NJ	3	3
6	Bristol	Bristol Borough	Bucks	PA	4	4
7	Burlington	Burlington City	Burlington	NJ	4	4
8	Beverly	Beverly	"	NJ	5	5
9	Delran	Delran	"	NJ	6	6
10	Riverside	Riverside	"	NJ	6	6
11	Palmyra	Palmyra	"	NJ	7	8
12	Riverton	Riverton	"	NJ	7	8
13	Cinnaminson	Cinnaminson Twp.	"	NJ	8	9
14	Philadelphia North East	Philadelphia	Philadelphia	PA	8	10
15	Pennsauken	Pennsuaken Twp.	Camdem	NJ	8	10
16	Camden Baldwin	Camden City	"	NJ	10	13
17	Camden Jackson	"	"	NJ	10	13
18	Philadelphia South East	Philadelphia	Philadelphia	PA	10	14
19	Gloucester City	Gloucester City	Camden	NJ	10	14
20	Brooklawn	Brooklawn Bor.	"	NJ	10	14
21	National Park	Nat'l Park Bor.	Gloucester	NJ	11	15
22	Paulsboro	Paulsboro Bor.	"	NJ	12	16
23	Philadelphia South West	Philadelphia	Philadelphia	PA	12	16
24	Greenwich	Greenwich Twp.	Gloucester	NJ	13	17
25	Tinicum	Tinicum Twp.	Delaware	PA	13	17
26	Eddystone	Eddystone Bor.	"	PA	13	17
27	Central Delaware County (Authority)	Ridley Twp.	"	PA	13	17
28	Chester	Chester City	"	PA	14	18
29	Marcus Hook	Marcus Hook Bor.	"	PA	15	19
30	Wilmington	Wilmington City	New Castle	DE	17	21
31	Penns Grove	Penns Grove Bor.	Salem	NJ	17	21
32	Upper Penns Neck	Up.Penns Nk Twp.	"	NJ	18	23
33	Pennsville	Pennsville Twp.	"	NJ	18	23
34	Delaware City	Del. City Twp.	New Castle	DE	20	27
35	Salem	Salem City	Salem	NJ	20	28
36	Lower Penns Neck	Pennsville	"	NJ	20	27
	Regional STP #1[a]	Ridley	Delaware	PA	13	17
	Regional STP #2[a]	Pennsville Twp.	Salem	NJ	18	23

[a]Do not exist in the region; information taken from Table 7.

Table B-1 (continued)

-- Part IV: Paper Plants and Paperstock Dealers (Module 5) --

No.	Name of Plant	Location						
		City	County	State	Coordinates[a]		Estuary reach[a]	
					x	y	RFF	DRBC
1	Scott Paper Co	Chester	Delaware	PA	469.6	4,410.3	14	18
2	Homasote	Trenton	Mercer	NJ	517.3	4,455.9	1	1
3	Nicolet Industries	Norristown	Montgomery	PA	470.9	4,440.0		
4	Linerboard[b]	Bristol	Bucks	PA	510.5	4,437.0	5	5
5	Newsprint[b]	Whitemarsh	Montgomery	PA	476.0	4,436.0	11	15
6	Georgia-Pacific[c]	Delair	Camden	NJ	495.0	4,425.0		
7	Sonoco	Downington	Chester	PA	440.0	4,428.9		
8	GAF Corporation	Gloucester City	Camden	NJ				
9	Whippany Paper Board	Riegelsville	Bucks	PA				
10	Brandywine et al	Downington	Chester	PA				
11	Celotex	Philadelphia	Philadelphia	PA	482.6	4,421.0		
12	Connelly Containers	"	"	PA	481.7	4,429.5		
13	Container Corporation of America	"	"	PA	479.4	4,431.6		
14	Crown Paper Board	"	"	PA				
15	Newman & Company	"	"	PA				
16	Paperstock Dealer #1[b]	Lester	Delaware	PA	476.0	4,414.0		
17	Paperstock Dealer #2[b]	Philadelphia	Philadelphia	PA	495.0	4,430.0		

[a] The identification of x-y coordinates and reach numbers in this table indicate residuals discharges in this module, except for paper stock dealers #1 and #2 which do not have gaseous discharges.

[b] Do not exist in the region.

[c] The liquid residuals discharges from, and residuals management options for, Georgia-Pacific are included in Module 6.

Table B-1 (continued)

-- Part V: Municipal Incinerators (Module 5) --

No.	Name of incinerator	Location						
		City	County	State	Coordinates		Estuary reach	
					x	y	RFF	DRBC
1	Philadelphia North East[a]	Philadelphia	Philadelphia	PA	493.2	4,425.9	8	10
2	Philadelphia East Central	"	"	PA	488.3	4,423.2	9	12
3	Philadelphia South East	"	"	PA	486.1	4,416.9	11	15
4	Philadelphia Harrowgate	"	"	PA	490.5	4,429.5		
5	Philadelphia Bartram	"	"	PA	481.9	4,420.6		
6	Philadelphia North West	"	"	PA	479.5	4,431.9		
7	Ewing Township	Ewing	Mercer	NJ	518.9	4,455.7		
8	Princeton Borough	Princeton	"	NJ	531.7	4,470.3		
9	Hamilton Townswhip	Hamilton	"	NJ	524.4	4,450.5		
10	Maple Shade Township[b]	Maple Shade	Burlington	NJ	500.3	4,423.6		
11	Moorestown Township[b]	Moorestown	"	NJ	510.7	4,427.3		
12	Chester Township	Chester	Delaware	PA	465.9	4,410.5		
13	Darby Township	Darby	"	PA	477.4	4,415.6		
14	Marple Township	Marple	"	PA	471.2	4,422.8		
15	Abington Township[a]	Abington	Montgomery	PA	489.7	4,439.6		
16	Lower Merion Township	Lower Merion	"	PA	479.3	4,430.7		
17	Whitemarsh Township	Whitemarsh	"	PA	477.0	4,436.0		
18	New Castle County[a]	Wilmington	New Castle	DE	455.0	4,396.0	17	22
19	Chester County[a]	East Pikeland	Chester	PA	455.0	4,445.0		
20	Bucks County[a]	Middletown	Bucks	PA	510.0	4,450.0		
21	Camden County[a]	Camden	Camden	NJ	490.0	4,418.0	10	13
22	Gloucester County[a]	Logan	Gloucester	NJ	470.0	4,409.0	14	18
23	Salem County[a]	Mannington	Salem	NJ	462.0	4,382.0		

[a]Did not exist in the region as of August 1973.

[b]Existing capacity only 25 tons per day.

Table B-1 (continued)

-- Part VI: Industrial Sewage Treatment Plants (Module 6) --

No.	Name of plant	Location				
		City	County	State	Estuary reach	
					RFF	DRBC
1	Patterson Parchment[a]	Bristol	Bucks	PA	3	3
2	Rohm & Haas[b]	"	"	PA	4	4
3	Hercules Powder[b]	Burlington	Burlington	NJ	4	4
4	Chamberlain[b]	"	"	NJ	5	5
5	Tenneco Plastics[a]	Burlington	Burlington	NJ	5	5
6	Georgia-Pacific[c]	Delair	Camden	NJ	8	10
7	National Sugar[a]	Philadelphia	Phil.	PA	9	11
8	MacAndrews-Forbes	Camden	Camden	NJ	10	13
9	Publicker	Philadelphia	Phil.	PA	10	13
10	Hershaw Chemical[a]	Gloucester City	Camden	NJ	10	14
11	GAF-Rubberoid[a]	"	"	NJ	10	14
12	NJ Zinc[b]	"	"	NJ	10	14
13	Philadelphia Naval Base[b]	Philadelphia	Phil.	PA	11	15
14	Shell Chemical[b]	Woodbury City	Gloucester	NJ	11	15
15	Hercules Powder[a]	Gibbstown	"	NJ	13	17
16	Dupont-Repauno[a]	"	"	NJ	13	17
17	FMC[b]	Marcus Hook Bor.	Delaware	PA	14	18
18	Monsanto[a]	Bridgeport	Gloucester	NJ	15	19
19	Allied Chemical[b]	Claymont	New Castle	DE	15	19
20	Dupont Edgemoor[b]	Edgemoor	"	DE	17	21
21	Dupont-Carney[b]	Penns Grove Bor.	Salem	NJ	17	21
22	Dupont-Chambers[b]	Deepwater	"	NJ	17	22
23	Atlas Chemical[b]	New Castle	New Castle	Del	17	22

[a]These eight activities have discharges of gaseous residuals in this module.

[b]The gaseous discharges from, and management options for, these 12 plants are included in Module 3.

[c]The gaseous discharges from, and management option for, this plant are included in Module 5.

Table B-2

Residuals Management Options Available to the Various Types of Dischargers in the Lower Delaware Valley Model

Type of residuals discharger	Management option available	Primary residual reduced	Secondary residual generated[a]
Petroleum refineries [1]	Change in raw material input mix:		
	1. Purchased fuel alternatives:		
	residual fuel oil--3 grades (sulfur contents: 0.5, 1.0, and 2.0 percent)	SO_2	None
	2. Charge of lower sulfur crude--2 grades (sulfur contents: 0.4 and 1.44 percent)	SO_2	None
	3. Sell, rather than burn, certain high-sulfur fuel products, e.g., sour refinery coke (3.32 percent sulfur); sweet coke (1.57 percent sulfur), desulfurized refinery gas (0.1 percent sulfur)	SO_2 Particulates	None
	Residuals modification processes:		
	1. Cyclone collectors on cat-cracker catalyst regenerator (two removal efficiencies: 70, 85 percent)	Particulates	Ash
	2. Electrostatic precipitator on cat-cracker catalyst regenerator (removal efficiency: 95 percent)	Particulates	Ash
	3. Secondary and tertiary treatment for sour water condensate	BOD, SS, N, toxics	Sludge
	4. Various reuse alternatives for treated wastewater (cooling tower water makeup, desalter water, boiler feedwater)	BOD	None
	5. Cooling tower(s) for segregated, noncontact cooling water	Heat	b
	6. Sludge drying and incineration	Sludge	Particulates Suspended solids

For footnotes, see page 167; for references, page 168.

continued ...

Table B-2 (continued)

Type of residuals discharger	Management option available	Primary residual reduced	Secondary residual generated[a]
Steel mills [2]	Change in raw material input mix:		
	1. Burden alternatives at blast furnace	c	
	2. Fuel alternatives:		
	a. residual fuel oil (0.5, 1.0, and 2.0 percent sulfur contents)	SO_2	
	b. natural gas	SO_2	
	Modification of production processes:		
	1. Scrap-hot metal ratio alternatives at open hearth furnace	c	
	Residuals modification processes:		
	1. Gas desulfurization at coke plant	SO_2	
	2. Cyclone collectors at sinter plant, blast furnace, and power plant (various removal efficiencies: from 60 to 98.5 percent)	Particulates	Ash
	3. Electrostatic precipitators at sinter plant, blast furnace, open hearth furnace, and basic oxygen furnace (various removal efficiencies: from 97 to 99.7 percent)	Particulates	Ash
	4. Fabric filter at sinter plant and electric arc furnace (various removal efficiencies: from 97 to 99.7 percent)	Particulates	Ash
	5. Wet scrubber at the blast furnace	Particulates	Ash Suspended solids
	6. Oil reclamation from cold mill emulsion	BOD	
	7. Recirculation of wastewater streams in hot rolling and cold mill finishing	BOD, SS, Oil, toxics, heat	

For footnotes, see page 167; for references, page 168.

continued ...

Table B-2 (continued)

Type of residuals discharger	Management option available	Primary residual reduced	Secondary residual generated[a]
Steel mills (continued)	8. Secondary treatment (coagulation, sand filtration, clarification and thickening, settling basins, lagoons, and activated sludge) in coke production, blast furnace, basic oxygen furnace, and finishing operations	BOD, SS	Sludge
	9. Tertiary treatment (ammonia stripping, and carbon adsorption) in coke production	BOD, N	
	10. Cooling towers in coke production, blast furnace, basic oxygen furnace, open hearth furnace, electric arc furnace, electric power plant, and finishing processes (removal efficiency: 100 percent)	Heat	b
	11. Sludge drying and incineration	Sludge	Particulates Suspended solids
Power plants [3]	1. Fuel alternatives:		
	a. coal--5 grades (sulfur contents: 0.5, 1.0, 1.5, 2.0, and 3.65 percent; ash contents: 7.4, 8.1, 8.8, 9.5, and 11.8 percent, respectively)	Particulates SO_2	
	b. residual fuel oil--5 grades (sulfur contents: 0.5, 1.0, 1.5, 2.0, and 3.65 percent)	Particulates SO_2	
	c. distillate fuel oil--3 grades (sulfur contents: 0.25, 0.55, and 1.35 percent)	Particulates SO_2	
	d. natural gas	Particulates SO_2	
	2. Catalytic oxidation--flue gas desulfurization	SO_2	
	3. Dry limestone injection	SO_2	Dry calcium sulfate
	4. Wet limestone injection	SO_2	Lime slurry

For footnotes, see page 167; for references, page 168.

continued ...

Table B-2 (continued)

Type of residuals discharger	Management option available	Primary residual reduced	Secondary residual generated[a]
Power plants (continued)	5. Centrifugal collector (three removal efficiencies: high, medium, and low)	Particulates	Bottom ash
	6. Electrostatic precipitator (three removal efficiencies: high, medium, and low)	Particulates	Bottom ash
	7. Fabric filter (two types: high and medium temperature)	Particulates	Bottom ash
	8. Cooling towers	Heat	b
Home heat [4]	Fuel alternatives:		
	1. coal (sulfur content: 1.5 percent; ash, 9.0 percent)		Bottom ash
	2. distillate fuel oil--2 grades (sulfur contents: 0.25 and 0.55 percent)	Particulates SO_2	
	3. residual fuel oil--5 grades (sulfur contents: 0.5, 1.0, 1.5, 2.0, and 2.5 percent)	Particulates SO_2	
	4. natural gas	Particulates SO_2	
Commercial heat [5]	Fuel alternatives:		
	1. residual fuel oil--5 grades (sulfur contents: 0.5, 1.0, 1.5, 2.0, and 2.5 percent)	SO_2	
	2. natural gas	Particulates SO_2	
Over 25 μgms/m^3 dischargers [6]	Management options for this group of dischargers are those provided in EPA's Implementation Planning Program (IPP) for air quality control [3]. The available options vary according to the particular industry and production process as represented by the 4-digit Standard Industrial Classification (SIC) code. For details, see Table 5-2 of Reference 3.	Particulates SO_2	Bottom ash SS

For footnotes, see page 167; for references, page 168.

continued ...

Table B-2 (continued)

Type of residuals discharger	Management option available	Primary residual reduced	Secondary residual generated[a]
Delaware Estuary sewage treatment plants [7]	1. Secondary wastewater treatment (activated sludge--removal efficiencies: BOD, 77; SS, 80, NH_3, 30; and PO_4, 22 percent)	BOD, SS, NH_3, PO_4	Sludge
	2. Tertiary wastewater treatment (combination of ammonia stripping, lime precipitation, and carbon adsorption--removal efficiencies: BOD, 67; SS, 99; NH_3, 88; and PO_4, 92 percent)	BOD, SS NH_3, PO_4	Sludge, Gas (e.g., NH_3)
	3. Sludge digestion, drying, and landfill	Sludge	d
	4. Raw sludge barged to sea	Sludge	d
	5. Sludge dewatering and incineration	Sludge	Particulates Bottom ash
	6. Dry cyclone on sludge incinerator (two removal efficiencies: 40 and 96 percent)	Particulates	Bottom ash
Paper plants [8]	1. Fuel alternatives: residual fuel oil--3 grades (sulfur contents: 0.5, 1.0, and 2.0 percent)	SO_2	
	2. Primary clarifier, aerated lagoon, and secondary clarifier (removal efficiencies: 30 combinations of BOD and SS removal)	BOD, SS	Sludge
Municipal incinerators	1. Wet baffles (two removal efficiencies: 70, 80 percent)	Particulates	SS, Bottom ash
	2. Flooded plate scrubber (removal efficiency: 90 percent)	Particulates	SS, Bottom ash
	3. Electrostatic precipitator (removal efficiency: 95 percent)	Particulates	Bottom ash
	4. Electrostatic precipitator plus high energy scrubber (removal efficiency: 98 percent)	Particulates	Bottom ash
	5. Settling ponds (removal efficiency: 90 percent)	SS	Sludge

For footnotes, see page 167; for references, page 168.

continued ...

Table B-2 (continued)

Type of residuals discharger	Management option available	Primary residual reduced	Secondary residual generated[a]
Municipal solid residuals handling and disposal	After collection:		
	1. Transport to, and disposal in, landfill	Solid residuals	e
	2. Transport to municipal incinerators plus landfill of incinerator residue	Solid residuals	Particulates Incinerator residue
	3. Transport to a transfer station and railhaul out of the region	Solid residuals	e
	4. Transport of newsprint to a processing center for reuse	Solid residuals	
Commercial corrugated board handling and disposal	After collection:		
	1. Transport to, and disposal in, landfill	Solid residuals	e
	2. Transport to a transfer station and railhaul out of the region	Solid residuals	e
	3. Transport to a processing center for reuse	Solid residuals	
Delaware Estuary industrial dischargers	Same management options employed in the Delaware Estuary sewage treatment plants module, above	BOD, SS, NH_3, PO_4, Toxics, Sludge	Sludge Gas (e.g., NH_3) d Particulates Bottom ash
		Particulates	Bottom ash

For footnotes, see page 167; for references, page 168.

continued ...

Footnotes to Table B-2:

[a]Excluding carbon dioxide and water vapor. Also, excluding any additional residuals generated in fuel combustion associated with the additional energy required for operating the residuals modification facilities. This latter consideration has, however, been incorporated in the regional model.

[b]Heat is rejected to the atmosphere along with water vapor.

[c]This alternative affects the quantities of liquid, gaseous, and solid residuals generated in production processes.

[d]The secondary residual here is digested sludge at a different location.

[e]The secondary residual here is municipal solid residuals at a different location.

References to Table B-2:

1. For details, see Clifford S. Russell, _Residuals Management in Industry: A Case Study of Petroleum Refining_ (Baltimore, Md.: The Johns Hopkins University Press for Resources for the Future, 1973).

2. For details, see Clifford S. Russell and William J. Vaughan, _Steel Production: Processes, Products, and Residuals_ (Baltimore, Md.: The Johns Hopkins University Press for Resources for the Future, 1976); and Clifford S. Russell and William J. Vaughan, "A Linear Programming Model of Residuals Management for Integrated Iron and Steel Production," _Journal of Environmental Economics and Management_, Vol. 1, No. 1 (1974), pp. 17-42.

3. For details, see TRW, Inc., "Air Quality Implementation Planning Program: Vol. 1, Operations Manual," U.S. Environmental Protection Agency, November 1970, Chapter 5. [Also available from the National Technical Information Service (NTIS), Springfield, Virginia 22161, accession no. PB-198 299.]

4. Taken from Walter O. Spofford, Jr., "Home Heating Module," RFF Lower Delaware Valley residuals management study, 5 Feb 1974 (unpublished).

5. Taken From Walter O. Spofford, Jr., "Commercial Heating Module," RFF Lower Delaware Valley residuals management study, 6 Feb 1974 (unpublished).

6. For details, see:

   (a) TRW, Inc., "Air Quality Implementation Planning Program: Vol. I, op. cit.

   (b) Clifford S. Russell and Marilyn E. McMillan, "Modification of the Implementation Planning Program," RFF Lower Delaware Valley residuals management study, no date (unpublished).

7. Taken from Clifford S. Russell, "The Regional Sewage Treatment Plant Module," RFF Lower Delaware Valley residuals management study, 31 Jan 1974 (unpublished). The cost functions for these alternatives were developed from information in R. Smith and R. Eilers, "Cost to the Consumer for Collection and Treatment of Wastewaters" (Washington, D.C.: U.S. Government Printing Office, July 1970).

8. Taken from James W. Sawyer, Jr., Blair T. Bower, and George O. G. Löf, "Modeling Process Substitutions by LP and MIP," in Robert M. Thrall et al., eds., _Economic Modeling for Water Policy Evaluation_, TIMS Studies in the Management Sciences, Vol. 3 (New York: North-Holland Publishing Co., 1976), pp. 157-178.

Table B-3

Housing Units Connected to Municipal Water and Sewerage Systems in Each Political Jurisdiction (1970)

Political Jurisdic-tion	State	County	Population	Housing units	Units connected to public:		Percent Housing units sewered
					water supply	sewers	
1	DE	New Castle	94,343	29,353	27,540	27,128	92.5
2	"	"	88,801	32,572	32,536	32,406	99.0
3	"	"	102,845	29,427	23,523	22,932	78.0
4	"	"	99,867	29,352	28,404	27,589	94.2
5	PA	Chester	87,052	24,331	17,515	12,265	50.5
6	"	"	83,550	24,949	17,007	12,716	51.0
7	"	"	81,550	23,874	10,790	8,408	35.2
8	PA	Delaware	99,733	33,909	33,682	33,737	99.4
9	"	"	102,256	31,624	31,575	31,620	100
10	"	"	104,061	31,294	31,095	30,998	99.0
11	"	"	108,395	34,296	33,964	33,450	97.4
12	"	"	91,918	26,290	20,958	16,780	63.9
13	"	"	93,672	27,027	26,424	24,385	90.0
14	PA	Philadelphia	90,688	29,689	29,670	29,558	99.6
15	"	"	103,102	41,753	41,716	41,494	99.5
16	"	"	93,930	35,744	35,665	35,434	99.0
17	"	"	92,656	28,827	28,817	28,474	98.8
18	"	"	99,495	30,904	30,857	30,698	99.3

continued ...

Source: Based on data obtained from the U.S. Bureau of the Census: <u>4th Count Housing Tapes</u>, 1970.

Table B-3 (continued)

Political Jurisdiction	State	County	Population	Housing units	Units connected to public: water supply	Units connected to public: sewers	Percent Housing units sewered
19	PA	Philadelphia (cont'd)	101,883	39,473	39,368	39,231	99.7
20	"	"	97,702	32,939	32,978	32,881	100
21	"	"	96,402	32,058	31,946	30,870	96.1
22	"	"	94,368	31,484	31,478	31,182	99.0
23	"	"	96,271	40,532	40,371	40,230	99.4
24	"	"	104,698	35,760	35,677	35,618	99.6
25	"	"	92,062	29,754	29,744	29,619	99.6
26	"	"	92,918	31,797	31,742	31,720	100
27	"	"	91,383	29,289	29,250	28,987	99.0
28	"	"	92,646	33,191	33,103	32,975	99.4
29	"	"	93,110	33,405	33,481	33,203	99.4
30	"	"	97,951	34,927	34,902	34,729	99.6
31	"	"	99,664	34,299	34,275	33,991	99.0
32	"	"	95,874	31,668	31,536	31,290	99.0
33	"	"	93,403	24,772	24,616	24,394	98.4
34	PA	Montgomery	100,066	32,753	32,349	29,409	89.7
35	"	"	103,520	31,656	31,064	29,800	94.2
36	"	"	100,813	28,646	19,286	19,988	69.5
37	"	"	111,691	36,153	35,737	33,798	93.5
38	"	"	108,663	32,389	27,653	24,092	74.3

continued ...

Table B-3 (continued)

Political Jurisdic-tion	State	County	Population	Housing units	Units connected to public:		Percent Housing units sewered
					water supply	sewers	
39	PA	Montgomery (cont'd)	99,046	31,995	19,167	20,539	64.1
40	PA	Bucks	112,621	33,338	30,183	31,278	93.8
41	"	"	103,815	29,451	26,241	26,137	89.0
42	"	"	101,298	27,474	15,621	18,600	67.9
43	"	"	97,322	31,447	11,484	11,403	35.3
44	NJ	Mercer	103,142	35,195	35,120	35,024	99.5
45	"	"	96,776	30,845	29,054	28,982	94.0
46	"	"	102,554	30,389	22,248	22,227	73.1
47	NJ	Burlington	92,251	27,581	25,045	24,373	88.0
48	"	"	91,931	25,640	24,241	24,074	94.0
49	"	"	91,811	28,455	18,514	16,653	58.6
50	NJ	Camden	95,717	32,018	31,909	31,602	99.0
51	"	"	105,214	30,985	30,308	29,848	96.5
52	"	"	86,342	29,827	29,742	29,575	99.2
53	"	"	84,653	25,261	23,356	23,964	95.0
54	"	"	77,531	22,664	17,341	14,260	62.9
55	NJ	Gloucester	87,767	26,060	21,977	18,796	72.0
56	"	"	84,914	25,648	15,798	13,854	54.0

continued ...

Table B-3 (continued)

Political Jurisdiction	State	County	Population	Housing units	Units connected to public:		Percent Housing units sewered
					water supply	sewers	
57	NJ	Salem	60,346	19,598	12,107	10,631	54.0
TOTALS			5,458,053	1,746,031		1,510,051[a]	86.5

[a]If the 11 political jurisdictions that do not discharge directly to the Delaware Estuary are excluded (nos. 5, 6, 7, 36, 39, 43, 45, 46, 53, 54, 56), the total number of housing units connected to public sewers would be 1,321,445.

Table B-4

Air Quality Receptor Locations
for the 57 Political Jurisdictions

Political jurisdiction	State	County	City/Township	Grid Coordinates	
				x	y
1	DE	New Castle	Brandywine Division	455	4405
2	"	"	Wilmington City	453	4400
3	"	"	Middletown-Odessa Div.	440	4375
4	"	"	New Castle Division	447	4390
5	PA	Chester	West Goshen Twp	450	4425
6	"	"	East Pikeland Twp	450	4440*
7	"	"	Newlin Twp	435	4420
8	PA	Delaware	Upper Darby Twp	476	4423
9	"	"	Collingdale Borough	476	4418
10	"	"	Morton Borough	472	4418
11	"	"	Nether Providence Twp	468	4415
12	"	"	Chester Heights Bor	460	4415
13	"	"	Radnor Township	470	4430*
14	PA	Philadelphia[a]	South Philadelphia	486.5	4419.5
15	"	"	Center City Phil.	488.0	4423.0
16	"	"	South Philadelphia	485.0	4420.5
17	"	"	South Philadelphia	482.0	4419.5
18	"	"	West Philadelphia	480.0	4422.0
19	"	"	West Philadelphia	483.0	4423.5
20	"	"	West Philadelphia	479.5	4425.0
21	"	"	Manayunk	482.0	4433.0
22	"	"	North Philadelphia	485.0	4426.5
23	"	"	North Philadelphia	486.0	4425.0
24	"	"	North Philadelphia	494.0	4429.0
25	"	"	Northeast (Frankford)	496.0	4432.0
26	"	"	North Philadelphia	491.5	4432.0

For footnotes, see page 175.

continued ...

Table B-4 (continued)

Political juris-diction	State	County	City/Township	Grid Coordinates	
				x	y
27	PA	Philadelphia (cont'd)	North Philadelphia	488.5	4427.0
28	"	"	North Philadelphia	486.0	4430.0
29	"	"	Germantown	484.0	4433.5
30	"	"	North Philadelphia	489.0	4431.0
31	"	"	Mount Airy	486.0	4435.0
32	"	"	Northeast (Frankford)	495.0	4438.0
33	"	"	Northeast (Frankford)	502.0	4439.0
34	PA	Montgomery	Lower Merion Twp	472	4435
35	"	"	Whitemarsh Twp	480	4440*
36	"	"	Worcester Twp	470	4450*
37	"	"	Abington Township	490	4440*
38	"	"	Upper Dublin Twp	485	4445
39	"	"	Lower Frederick Twp	460	4460*
40	PA	Bucks	Bristol Township	510	4440*
41	"	"	Falls Township	515	4449
42	"	"	Northampton Twp	500	4450*
43	"	"	Hilltown Township	480	4470*
44	NJ	Mercer	Trenton City	520	4452
45	"	"	Hamilton Twp	530	4450*
46	"	"	Lawrence Twp	526	4463
47	NJ	Burlington	Moorestown Twp	505	4425
48	"	"	Burlington Twp	510	4433
49	"	"	Southampton Twp	520	4420*
50	NJ	Camden	Camden City	490	4420*
51	"	"	Cherry Hill Twp	497	4420
52	"	"	Oaklyn Borough	493	4417
53	"	"	Gloucester Twp	495	4410
54	"	"	Waterford Twp	510	4400*

For footnotes, see page 175.

continued ...

Table B-4 (continued)

Political juris-diction	State	County	City/Township	Grid Coordinates	
				x	y
55	NJ	Gloucester	East Greenwich Twp	480	4405
56	"	"	Glassboro Borough	490	4395
57	NJ	Salem	Mannington Twp	470	4385

*Located on regularly spaced 10 kilometer grid.

[a]West Philadelphia is defined as the area west of the Schuylkill River.

South Philadelphia is defined as the area east of the Schuylkill River and south of South Street.

Center City is defined as the area east of the Schuylkill River, north of South Street, and south of Spring Garden Street.

North Philadelphia is defined as the area north of Spring Garden Street and east of Broad Street up to the area known as Northeast Philadelphia or Frankford.

Northeast Philadelphia (Frankford) is defined as the area northeast of Oxford and Cheltenham Avenues and Van Kirk Street.

Manayunk is defined as the area in Northwest Philadelphia between Fairmount Park and the Schuylkill River.

Germantown is defined as the area in Northwest Philadelphia surrounding the intersection of Washington Lane and Germantown Avenue.

Mount Airy is defined as the area in Northwest Philadelphia bordering Montgomery County in the vicinity of Mount Airy Avenue.

Appendix C

Supplemental Output of Production Runs of the Lower Delaware Valley Model

Contents

Table C-1

Distribution of Increased Consumer Costs by Political Jurisdiction for Selected Production Runs

Political jurisdiction[a]	Home heating (percent)		Household electricity (percent)		Municipal solid waste (percent)		Municipal sewage ($/household/year)	
	Run 1[b]	Run 3	Run 1	Run 3	Run 3	Run 7	Run 1	Run 2
1	25.1	1.8	15.8**	2.4	19.2	7.7	0.5*	0.1
2	30.5	0.7	15.8**	2.4	19.4	7.8	43.5	8.0
3	29.0	0*	15.8**	2.4	19.3	7.7	5.9	1.1
4	23.0	0*	15.8**	2.4	19.4	7.8	1.0	0.2
5	33.9	0.6	13.6	4.5**	19.1	7.6	n.a.	n.a.
6	35.5	0*	13.6	4.5	19.1	7.7	n.a.	n.a.
7	16.8	0*	13.6	4.5	19.4	7.8	n.a.	n.a.
8	26.1	9.6	13.6	4.5	10.8	7.7	3.3	2.5
9	23.3	7.2	13.6	4.5	17.9	7.7	2.1	1.0
10	24.5	3.2	13.6	4.5	19.3	7.7	3.5	1.9
11	28.3	0.9	13.6	4.5	10.9	7.7	16.2	8.0
12	27.8	3.4	13.6	4.5	18.8	7.7	3.9	2.1
13	26.2	1.9	13.6	4.5	21.0	7.7	2.5	1.7
14	19.7	6.1	13.6	4.5	8.7	7.7	21.4	16.1
15	41.0	25.8	13.6	4.5	7.6*	7.6	16.2	12.2
16	31.4	18.5	13.6	4.5	17.3	7.7	15.4	10.6
17	16.4	7.4	13.6	4.5	8.1	7.7	19.3	13.1
18	21.7	8.2	13.6	4.5	17.5	7.7	18.8	12.8
19	50.4	28.5	13.6	4.5	20.1	7.6	14.4	9.8
20	22.2	6.3	13.6	4.5	20.4	7.7	16.8	11.4
21	25.6	6.3	13.6	4.5	8.6	7.6	16.6	11.6
22	31.7	15.2	13.6	4.5	20.4**	7.7	19.3	14.0
23	55.4**	30.0**	13.6	4.5	11.2	7.5	14.8	10.6
24	14.1	0.2	13.6	4.5	17.6	6.6*	8.0	7.1
25	10.4	0.5	13.6	4.5	17.8	6.6*	16.0	14.2
26	12.6	0.9	13.6	4.5	23.1	8.2**	15.0	13.4

For footnotes, see page 179.

continued ...

Table C-1 (continued)

Political jurisdiction[a]	Home heating (percent)		Household electricity (percent)		Municipal solid waste (percent)		Municipal sewage ($/household/year)	
	Run 1[b]	Run 3	Run 1	Run 3	Run 3	Run 7	Run 1	Run 2
27	25.5	4.8	13.6	4.5	20.5	7.7	22.1	16.7
28	29.8	5.1	13.6	4.5	20.4	7.7	15.6	13.3
29	33.9	1.6	13.6	4.5	10.7	7.6	17.9	14.1
30	16.5	2.7	13.6	4.5	20.6	7.7	15.6	13.9
31	17.2	2.7	13.6	4.5	20.3	7.7	15.7	13.6
32	9.9*	0.6	13.6	4.5	20.2	7.6	15.8	14.0
33	31.4	1.1	13.6	4.5	20.4	7.7	19.7	17.6
34	35.1	1.9	13.6	4.5	18.7	7.5	7.3	4.9
35	26.9	2.8	13.6	4.5	19.1	7.7	2.3	1.5
36	31.2	0.0	13.6	4.5	19.2	7.7	n.a.	n.a.
37	26.7	0.7	13.6	4.5	19.0	7.6	6.1	5.4
38	26.9	0.6	13.6	4.5	19.0	7.6	1.4	1.2
39	36.7	0*	n.a.	n.a.	19.2	7.7	n.a.	n.a.
40	36.7	0.2	13.6	4.5	19.1	7.6	8.3	1.3
41	31.3	0*	13.6	4.5	19.2	7.7	15.9	0.0*
42	29.6	0.3	13.6	4.5	19.3	7.7	1.8	1.6
43	33.1	0*	6.8	2.2	19.3	7.7	n.a.	n.a.
44	34.9	0*	2.9	0.4	19.4	7.7	48.2**	1.2
45	24.2	0*	2.9	0.4	19.2	7.7	n.a.	n.a.
46	25.0	0*	2.9	0.4	19.2	7.7	n.a.	n.a.
47	25.1	0.8	2.9	0.4	19.3	7.7	12.7	3.0
48	17.4	0*	2.9	0.4	19.5	7.8	11.5	0.0*
49	19.7	0*	1.4*	0.2*	19.4	7.8	2.9	0.0*
50	31.7	9.2	2.9	0.4	19.5	7.8	47.8	36.9**
51	20.6	4.1	2.9	0.4	19.3	7.7	20.7	13.6
52	29.5	6.0	2.9	0.4	19.4	7.8	4.4	3.2
53	22.3	2.8	2.9	0.4	19.4	7.8	n.a.	n.a.
54	25.6	0.7	10.6	1.3	19.3	7.7	n.a.	n.a.

For footnotes, see page 179.

continued ...

Table C-1 (continued)

Political jurisdiction[a]	Home heating (percent)		Household electricity (percent)		Municipal solid waste (percent)		Municipal sewage ($/household/year)	
	Run 1[b]	Run 3	Run 1	Run 3	Run 3	Run 7	Run 1	Run 2
55	25.9	5.0	2.9	0.4	19.5	7.8	6.6	4.5
56	19.2	0.2	10.6	1.3	19.5	7.8	n.a.	n.a.
57	28.2	0*	10.6	1.3	19.6	7.9	0.5	0.1

[a]For the state and county locations of the 57 political jurisdictions, see Tables B-3 and B-4 in Appendix B.

[b]Cost increases for home heating are all at the technological upper limits except for jurisdictions 1, 2, 7, 12, 35, 49, and 56. There is empirical evidence to suggest that at the "optimum" for this run, all cost increases would be at their upper limits. Cost increases at the upper limits represent a situation where there is total conversion within the region to natural gas.

* Minimum increases

** Maximum increases

n.a. = not applicable

Table C-2

Distribution of Air Quality by Political Jurisdictions
for Selected Production Runs
-- Sulfur Dioxide Concentrations --

Political Jurisdiction[a]	Sulfur dioxide (μgms/m^3)		Political Jurisdiction[a]	Sulfur dioxide (μgms/m^3)	
	Run 1	Run 3		Run 1	Run 3
1	17	54	30	23	66
2	15	49	31	16	46
3	6	14	32	13	44
4	12	39	33	12	40
5	8	22	34	11	35
6	5	19	35	11	32
7	5	12	36	6	20
8	22	53	37	12	37
9	32*	68	38	9	27
10	25	58	39	4	12
11	28	64	40	11	46
12	15	42	41	9	34
13	11	32	42	7	24
14	56*	96*	43	4	11
15	42*	98*	44	13	40
16	83*#	120*#	45	10	34
17	64*	100*	46	8	20
18	35*	70	47	12	49
19	50*	91*	48	11	40
20	24	57	49	8	25
21	18	48	50	40*	87*
22	29	74*	51	20	64
23	38*	89*	52	27	71
24	29	103*	53	16	43
25	19	66	54	6	18
26	25	70	55	22	51
27	34*	94*	56	9	26
28	25	65	57	8	26
29	17	49			

[a]For the state and county locations of the 57 political jurisdictions, see Tables B-3 and B-4 in Appendix B.

*Indicates the 10 highest concentrations of sulfur dioxide.

#Indicates the maximum concentration of sulfur dioxide.

Table C-3

Distribution of Water Quality by Estuary Reach for Selected Production Runs -- Dissolved Oxygen --

Delaware Estuary Reach[a]	Dissolved oxygen (mg/ℓ)					
	Run 1	Run 2	Run 3	Run 13	Run 15	Run 16
1	8.6	8.6	8.6	8.6	8.6	8.6
2	7.7	7.7	8.0	7.9	7.9	7.9
3	6.9	6.6	7.2	7.0	6.9	7.2
4	6.3	6.0	6.8	6.2	6.1	6.5
5	5.9*	5.7	6.4	5.8*	5.4*	6.0
6	6.0	5.9	6.5	6.0	5.4*	6.0
7	5.9*	5.9	6.3	5.8*	5.3*	6.0
8	5.9	5.8	6.3	6.2	5.7	5.8*
9	6.4	6.1	6.7	6.8	6.3	6.2
10	6.8	5.3	7.0	7.8	7.5	7.1
11	5.3*	3.6*	5.1*	5.7*	5.2*	5.1*
12	6.1	3.7*	5.7*	6.3	6.0	6.1
13	5.7*	3.6*	5.4*	5.7*	5.3*	5.0*
14	5.7*	4.0*	5.5*	5.7*	5.3*	5.1*
15	6.1	4.5*	6.0*	6.0	5.5*	5.8*
16	6.4	5.2	6.3	6.0*	5.6	6.4
17	5.0*#	3.0*#	5.0*#	4.9*#	5.0*#	5.0*#
18	5.1*	3.4*	5.2*	5.4*	5.5	5.1*
19	5.7*	4.8*	5.8*	6.0	6.1	5.8*
20	6.1	5.8	6.1	6.1	6.2	6.1
21	6.2	6.2	6.3	6.3	6.3	6.3
22	6.6	6.6	6.6	6.7	6.7	6.6

[a]For locations of reaches, see Figure 5.

*Indicates the 8 lowest concentrations of dissolved oxygen.

#Indicates the minimum concentration of dissolved oxygen

References

1. Clifford S. Russell and Walter O. Spofford, Jr., "A Quantitative Framework for Residuals-Environmental Quality Management," in Gerrit H. Toebes, ed., Natural Resources Systems Models in Decision Making, Water Resources Research Center, Purdue University, 1969.

2. Clifford S. Russell and Walter O. Spofford, Jr., "A Quantitative Framework for Residuals Management Decisions," in Allen V. Kneese and Blair T. Bower, eds., Environmental Quality Analysis: Theory and Method in the Social Sciences (Baltimore, Md.: The Johns Hopkins University Press for Resources for the Future, 1972).

3. Clifford S. Russell, Walter O. Spofford, Jr., and Edwin T. Haefele, "The Management of the Quality of the Environment," in Jerome Rothenberg and Ian G. Heggie, eds., The Management of Water Quality and the Environment (New York: Halsted Press, 1974).

4. Walter O. Spofford, Jr., Clifford S. Russell, and Robert A. Kelly, "Operational Problems in Large-Scale Residuals Management Models," in Edwin S. Mills, ed., Economic Analysis of Environmental Problems (New York: National Bureau of Economic Research, 1975).

5. Walter O. Spofford, Jr., "Total Environmental Quality Management Models," in Rolf A. Deininger, ed., Models for Environmental Pollution Control (Ann Arbor, Mich.: Ann Arbor Science Publishers, Inc., 1973). [Available as Reprint No. 130, "Integrated Residuals Management: A Regional Environmental Quality Management Model," Resources for the Future, Inc., 1976.]

6. Robert A. Kelly, "Conceptual Ecological Model of the Delaware Estuary," in Bernard C. Patten, ed., Systems Analysis and Simulation in Ecology, Vol. IV (New York: Academic Press, forthcoming).

7. Robert A. Kelly and Walter O. Spofford, Jr., "Application of an Ecosystem Model to Water Quality Management: The Delaware Estuary," in Charles A. S. Hall and John W. Day, Jr., eds., Models as Ecological Tools: Theory and Case Histories (New York: Wiley-Interscience, forthcoming).

8. Clifford S. Russell, Walter O. Spofford, Jr., and Robert A. Kelly, "Early Returns on the Prospects for Regional Residuals Management Models," presented at the National Meeting of the Institute of Management Science/Operations Research Society of America, Boston, Massachusetts, April 1974.

9. Robert A. Kelly, "The Delaware Estuary," in Clifford S. Russell, ed., Ecological Modeling in a Resource Management Framework (Washington, D.C.: Resources for the Future, Inc., 1975).

10. Walter O. Spofford, Jr., Clifford S. Russell, and Robert A. Kelly, "The Delaware Valley Integrated Residuals Management Model: A Summary and Some Results," presented at Conference on Regional Residuals-Environmental Quality Management Models, sponsored by the World Health Organization and Resources for the Future, Inc., Rotterdam, 22-25 October 1974.

11. Clifford S. Russell, Walter O. Spofford, Jr., and Robert A. Kelly, "Interdependencies Among Gaseous, Liquid, and Solid Residuals: The Case of the Lower Delaware Valley," The Northeast Regional Science Review, Vol. 5 (Nov., 1975).

12. Clifford S. Russell, "Models for the Investigation of Industrial Response to Residuals Management Action," The Swedish Journal of Economics, Vol. 73, No. 1 (March 1971). [Available as Reprint No. 95, Resources for the Future, Inc., June 1971.]

13. Clifford S. Russell, Residuals Management in Industry: A Case Study of Petroleum Refining (Baltimore, Md.: The Johns Hopkins University Press for Resources for the Future, 1973).

14. Clifford S. Russell and William J. Vaughan, "A Linear Programming Model of Residuals Management for Integrated Iron and Steel Production," Journal of Environmental Economics and Management, Vol. 1, No. 1 (May 1974), pp. 17-42. [Available as Reprint no. 114, Resources for the Future, Inc., July 1974.]

15. Clifford S. Russell and William J. Vaughan, Steel Production: Processes, Products, and Residuals (Baltimore, Md.: The Johns Hopkins University Press for Resources for the Future, Inc., 1976).

16. Blair T. Bower, George O. G. Löf, and W. M. Hearon, "Residuals Management in the Pulp and Paper Industry," Natural Resources Journal, Vol. 11, No. 4 (October 1971), pp. 605-623. [Available as Reprint no. 100, Resources for the Future, Inc., 1971.]

17. George O. G. Löf, W. M. Hearon, and Blair T. Bower, "Residuals Management in Pulp and Paper Manufacture," in Walter S. Kaghan, ed., Forest Products and the Environment, AIChE Symposium Series No. 133, Vol. 69 (1973) pp. 141-149.

18. James W. Sawyer, Jr., Blair T. Bower, and George O. G. Löf, "Modeling Process Substitutions by LP and MIP," in Robert M. Thrall et al., eds., Economic Modeling for Water Policy Evaluation, TIMS Studies in the Management Sciences, Vol. 3 (New York: North-Holland Publishing Co., 1976), pp. 157-178.

19. Walter O. Spofford, Jr., "Solid Residuals Management: Some Economic Considerations," Natural Resources Journal, Vol. 11, (1971), pp. 561-89. [Available as Reprint no. 98, Resources for the Future, Inc., October 1971].

20. U.S. Department of Commerce, Bureau of the Census, 1970 Census of Population: Pennsylvania, PC(VI)-40, January 1971.

21. U.S. Department of Commerce, Bureau of the Census, 1970 Census of Population: New Jersey, PC(VI)-32, January 1971.

22. U.S. Department of Commerce, Bureau of the Census, 1970 Census of Population: Delaware, PC(VI)-9, January 1971.

23. U.S. Department of Commerce, Bureau of the Census, Census Tracts: Philadelphia, PA-NJ SMSA, Report PHC(1)-159, May 1972.

24. U.S. Department of Commerce, Bureau of the Census, Census Tracts: Trenton, N.J. SMSA, Report PHC(1)-217, April 1972.

25. U.S. Department of Commerce, Bureau of the Census, Census Tracts: Wilmington, DEL-NJ-MD SMSA, Report PHC(1)-234, March 1972.

26. U.S. Department of Commerce, Bureau of the Census, Area Measurement Reports: Areas of Pennsylvania, 1960, GE-20, No. 40, November 1967.

27. U.S. Department of Commerce, Bureau of the Census, Area Measurement Reports: Areas of New Jersey, 1960, GE-20, No. 32, January 1967 (revised).

28. U.S. Department of Commerce, Bureau of the Census, Area Measure-Reports: Areas of Delaware and District of Columbia, 1960, GE-20, Nos. 9 and 10, February 1967.

29. The World Almanac and Book of Facts, Newspaper Enterprise Association, Inc., New York City, 1973.

30. U.S. Geological Survey, Surface Water Supply of the United States: Part I -- North Atlantic Slope Basins, through 1950, Geological Survey Water Supply Paper 1302; 1950-1960, Paper 1722; 1960-1965, Paper 1902 (U.S. Government Printing Office, Washington, D.C., 1960, 1964, and 1970, respectively).

31. H. C. Wohlers and W. E. Jackson, "Air Pollution Emissions in the Delaware Valley for 1965," Drexel Institute of Technology, Philadelphia, June 1968 (prepared for the Regional Conference of Elected Officials).

32. J. Korshover, "Climatology of Stagnating Anticyclones East of the Rocky Mountains, 1936-1965," Public Health Service, USHEW, Publication No. 999-AP-34, 1967.

33. F. K. Davis, "The Air Over Philadelphia," Symposium: Air Over Cities, U.S. Public Health Service, Technical Report A62-5, November 1961, pp. 115-129.

34. J. R. Mather, "Meteorology and Air Pollution in the Delaware Valley," C. W. Thornthwaite Associates, November 1967, Table 2 (prepared for the Regional Conference of Elected Officials).

35. G. C. Holzworth, "Estimates of Mean Maximum Mixing Depths in the Contiguous United States," Monthly Weather Review, Vol. 92, No. 5 (May 1964), pp. 235-242; Figures 2-13, p. 238-239.

36. U.S. Department of Commerce (ESSA), "Climatic Atlas of the United States" (Washington, D.C.: U.S. Government Printing Office, June 1968).

37. Delaware River Basin Commission, "Final Progress Report: Delaware Estuary and Bay Water Quality Sampling and Mathematical Modeling Project," May 1970, Figure 12.

38. G. A. Rohlich and O. D. Uttormark, "Wastewater Treatment and Eutrophication," Nutrients and Eutrophication, Special Symposium Volume 1, American Society of Limnology and Oceanography, Inc., 1972, pp. 231-45.

39. Robert V. Thomann, Systems Analysis and Water Quality Management, Environmental Science Services Division of Environmental Research and Applications, Inc., New York, 1972.

40. U.S. Environmental Protection Agency, Storm Water Management Model, (Washington, D.C.: U.S. Government Printing Office, 1971), p. 180.

41. U.S. Environmental Protection Agency, "Application of Implementation Planning Program (IPP) Modeling Analysis: Metropolitan Philadelphia Interstate AQCR," Air Quality Management Branch, USEPA, Durham, North Carolina, February 1972 (unpublished).

42. Anthony V. Fiacco and Garth P. McCormick, Nonlinear Programming: Sequential Unconstrained Minimization Techniques (New York: John Wiley & Sons, Inc., 1968).

43. W. I. Zangwill, Nonlinear Programming: A Unified Approach (Englewood Cliffs, N.J.: Prentice-Hall, Inc., 1969).

44. TRW, Inc., "Air Quality Implementation Planning Program," Vols. I and II, November 1970 (prepared by TRW, Inc. for the U.S. Environmental Protection Agency). [Also available from National Technical Information Service, Springfield, Virginia 22161, Nos. PB-198 299 and PB-198 300, respectively.]

45. Walter O. Spofford, Jr., "Township and Jurisdiction Codes," RFF Lower Delaware Valley residuals management study, 17 July 1973 (unpublished).

46. G. Ozolins and R. Smith, "A Rapid Survey Technique for Estimating Community Air Pollution Emissions," Public Health Service, Publication No. 999-AP-29, National Air Pollution Control Administration (HEW), Raleigh, North Carolina, October 1968.

47. Walter O. Spofford, Jr., "Home Heating Module," RFF Lower Delaware Valley residuals management study, 5 February 1974 (unpublished).

48. Walter O. Spofford, Jr., "Commercial Heating Module," RFF Lower Delaware Valley residuals management study, 6 February 1974 (unpublished).

49. Clifford S. Russell and Marilyn E. McMillan, "Modification of the Implementation Planning Program," RFF Lower Delaware Valley residuals management study, no date (unpublished).

50. R. Smith and R. Eilers, "Cost to the Consumer for Collection and Treatment of Wastewaters" (Washington, D.C.: U.S. Government Printing Office, July 1970).

51. Elbert E. Whitlatch, Jr., "Optimal Siting of Regional Wastewater Treatment Plants," Ph.D. thesis, The Johns Hopkins University, Baltimore, Maryland, 1973.

52. William Whipple, Jr. and others, "Instream Aeration of Polluted Rivers," Water Resources Research Institute, Rutgers University, 1969.

53. William Whipple, Jr. et al., "Oxygen Regeneration of Polluted Rivers: The Delaware River," Research Report No. 16080, Water Quality Office, U.S. Environmental Protection Agency, December 1970.

54. William Whipple, Jr. and Shaw L. Yu, "Alternative Oxygenation Possibilities for Large Polluted Rivers," Water Resources Research, Vol. 7, No. 3 (June 1971), pp. 566-579.

55. Walter O. Spofford, Jr., "Instream Aeration," RFF Lower Delaware Valley residuals management study, 6 November 1973 (unpublished).

56. Walter O. Spofford, Jr., "Stack Aggregation for First-Stage Delaware Valley Model," RFF Lower Delaware Valley residuals management study, 5 September 1973 (unpublished).

57. Walter O. Spofford, Jr., "Relationship between Political Jurisdictions and Electric Company Service Areas," RFF Lower Delaware Valley residuals management study, 26 September 1973 (unpublished).

58. U.S. Environmental Protection Agency, "Implementation Plan Review for Pennsylvania as Required by the Energy Supply and Environmental Coordination Act," Report No. EPA-450/3-75-019, Office of Air Quality Planning and Standards, Research Triangle Park, N.C., February 1975.

59. U.S. Environmental Protection Agency, Monitoring and Air Quality Trends Report, 1974, Report No. EPA-450/1-76-001, Office of Air Quality Planning and Standards, Research Triangle Park, North Carolina, February 1976.

60. "Control District News," Journal of the Air Pollution Control Association, Vol. 24, No. 12 (December 1974) pp. 1202-1203.

61. Walter O. Spofford, Jr., "Background Concentrations of Suspended Particulates Due to Area Sources," RFF Lower Delaware Valley residuals management study, 13 February 1974 (unpublished), Table 2.

62. Walter O. Spofford, Jr., "Ecological Modeling in a Resource Management Framework: An Introduction" in Clifford S. Russell, ed., Ecological Modeling in a Resource Management Framework (Washington, D.C.: Resources for the Future, 1975), chapter 1.

63. Clifford S. Russell, ed., Ecological Modeling in a Resource Management Framework (Washington, D.C.: Resources for the Future, 1975).

64. Blair T. Bower and Ann M. Blackburn, "Dollars and Sense," a Report on Water Quality Management in the Washington Metropolitan Area, Interstate Commission on the Potomac River Basin, Bethesda, Maryland, September 1975.

65. Hydroscience, Inc., "Chesapeake Bay Wasteload Allocation Study" (prepared for the Maryland Department of Natural Resources), Westwood, New Jersey, April 1975.

66. Blair T. Bower, "The Economics of Industrial Water Utilization," in Allen V. Kneese and Stephen C. Smith, eds., Water Research (Baltimore: The Johns Hopkins University Press for Resources for the Future, Inc., 1966).

67. Blair T. Bower, G. P. Larson, A. Michaels, and W. M. Phillips, Waste Management: Generation and Disposal of Solid, Liquid, and Gaseous Wastes in the New York Region, Regional Plan Association, New York, 1968.

68. Federal Water Pollution Control Administration, "Delaware Estuary Comprehensive Study," (Philadelphia, Pa.: U.S. Department of the Interior, July 1966).

69. James A. Calloway, Andrew K. Schwartz, Jr., and Russell G. Thompson, "Industrial Economic Model of Water Use and Waste Treatment for Ammonia," Water Resources Research, Vol. 10, No. 4, pp. 650-658, August 1974.

70. F. Dail Singleton, Jr., James A. Calloway, and Russell G. Thompson, "An Integrated Power Process Model of Water Use and Wastewater Treatment in Chlor-Alkali Production," Water Resources Research, Vol. 11, No. 4, pp. 515-525, August 1975.

71. John C. Stone, Andrew K. Schwartz, Jr., Robert A. Klein, F. Dail Singleton, Jr., James A. Calloway, and Russell G. Thompson, "An Integrated Power Process Model of Water Use and Waste Treatment in Ethylene Production," Water Resources Research, Vol. 11, pp. 810-814, December 1975.